PRANK LAB

25 HILARIOUS EXPERIMENTS*

*SCIENTIFIC PRACTICAL JOKES

WADE DAVID FAIRCLOUGH + CHRIS FERRIE + BYRNE LAGINESTRA

 sourcebooks
eXplore

For my beautiful daughter, Avalon.

To the Fam Bam.

To Lindsay. It wasn't me.

Text © 2021 Wade David Fairclough, Chris Ferrie, Byrne LaGinestra
Illustrations © Wade David Fairclough
Cover and internal design © 2021 by Sourcebooks
Cover design by Maryn Arreguín/Sourcebooks
Internal design by Jillian Rahn/Sourcebooks
Internal images © Cat_Chat/Getty Images, CSA-Archive/Getty Images, dooder/Freepik.com, doodlebarn/Freepik.com, Freepik.com, Natalie_/Getty Images, olllikeballoon/Freepik.com, ourlifelooklikeballoon/Getty Images, Rakdee/Getty Images, rawpixel.com/Freepik.com, Sashatigar/Getty Images, shtonado/Getty Images, starline/Freepik.com, tartila/Freepik.com, Tetiana Lazunova/Getty Images, tovovan/Freepik.com, user13413638/Freepik.com, user6940940/Freepik.com, Valencya Tanaya/Getty Images, vectortatu/Getty Images

Published by Sourcebooks eXplore, an imprint of Sourcebooks Kids
P.O. Box 4410, Naperville, Illinois 60567-4410
(630) 961-3900
sourcebookskids.com

Library of Congress Cataloging-in-Publication Data is on file with the publisher.

Source of Production: 1010 Printing Asia Limited, Kwun Tong, Hong Kong, China
Date of Production: April 2022
Run Number: 5026209

Printed and bound in China.
OGP 10 9 8 7 6 5 4 3 2

CONTENTS

INTRODUCTION

So you've decided you want to pull some pranks, huh? You probably want to get straight into tormenting your sister, or brother, or whoever your target is. But skipping this introduction would be a mistake.

First, make sure you have a special place for this book, because before long your friends and family might be looking to trash it. Second, you need to understand that pranking is a noble cause; it enlightens the heart and brings joy to people, especially yourself. The Hall of Fame for pranksters includes writers, artists, inventors, even kings and queens! Though it is hard to say when the first-ever practical joke happened, it probably happened the same day that humans invented comedy. We have some pretty good evidence for jokers over fifteen hundred years ago, including a joke bowl that would pour its contents all over the victim trying to take a sip out of it!

Pulling off the perfect prank isn't just about following some instructions. NO! It takes planning, it takes patience, and perhaps most of all, it takes a keen eye to know when someone is READY

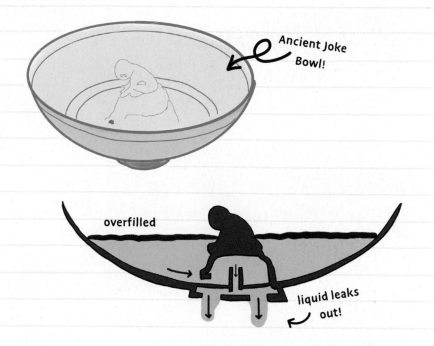

Ancient Joke Bowl!

overfilled

liquid leaks out!

TO BE PRANKED. You don't want to get all *Mentos Mayhem* on your buddy only to find out they were wearing their expensive clothes, or laugh at your mom's *ArachNOphobia* when she's spent all afternoon cleaning up after you—she probably won't see the funny side. To pull off the perfect prank, you need to be sure your victim isn't expecting it, but that they are also ready for it. That way, everyone gets to enjoy the joke, and you are less likely to be grounded for the rest of your life.

Some of the pranks in this book are VERY messy, so you will need to plan **where** and **when** you are going to unleash them. Choose somewhere easy to clean up, as it is very likely that you will be doing the cleaning. You will also need the supplies to clean up with (mostly towels).

All of the pranks in this book require stuff you can find in pretty much every house, but you'll still need to check if you have all the "ingredients" before starting any prank. Remember, it may seem a little weird if you keep asking your parents to get you some food coloring, baking soda, or bitter-tasting nail polish. Similarly, if you are continually carrying a towel before giving people food and drinks, they'll start to think something is up. To keep the suspicions low, you should check for ingredients for all of the pranks and note down what's missing. Then you can stock up on anything else you don't have in one shopping swoop!

So, now you have the perfect victim, place, and all the ingredients you need. No one suspects a thing. You have been ultra careful to read the instructions precisely, and you have perfectly organized how you will deal with the aftermath of the prank. Let's do this!

NO! Wait!

You need to understand each prank in this book will likely work slightly differently from what the instructions say. People think pranking is an art, where you put everything in place and BAM, you produce a piece of work that will be remembered for all time! In reality, pranking is more like science. In fact, one of history's most famous scientists was a devoted prankster. Benjamin Franklin—known for flying kites in thunderstorms and arguing with Britons—LOVED to pull pranks. He often wrote fake articles to newspapers pretending to be other people, one time even convincing the public another scientist was dead and that it was an impostor pretending to do his work. Just like scientists, pranksters need to be observant and should continually fine-tune their work to ensure the desired results, whether it be the embarrassment of your friends or having your sister clean your room for a month. Each time you pull a prank, you get better and better and better, and pretty soon, you will able to predict what will happen 95 percent of the time.

This is why we included a "Prank Review" section for you to use after each prank. Whenever you pull a fast one, go back and answer the questions in that section. Before long, you'll be a professor-level prankster!

MAKING A MESS

The practical jokes in this section can stain, soak, or smear themselves on your victims, furniture, or walls. When pulling a messy prank, you need to start at the end and make sure you have cleaning gear at the ready, but keep it hidden away so your victim doesn't suspect anything. You should choose the best conditions for the prank, because it can be hard to clean chocolate from a wedding dress or soda off a roof, and carpet can take a long time to dry.

To maximize the funniness of these pranks, you will need to know some chemistry as well as some physics. But don't worry, we've got all the tips and tricks you'll need to get started on your own pranking processes. Don't forget to try the "Prank Review" to perfect your skills.

EXPERIMENT #1

FOUNTAIN DEW

VICTIM: Dad

FUNNINESS: 8

MESS: 7

SCIENCE: 10

DANGER: 3

SHOPPING LIST

☐ Plastic bottle (with lid!)

☐ Water

☐ Food coloring

☐ Pin

☐ Towel

WARNINGS TO FUTURE ME

- Don't do this in the bedroom again!
- Best to do this prank in the kitchen or outside.
- Play this prank when Dad is not wearing his work clothes!

PRANK PROCEDURE: FOUNTAIN DEW

INSTRUCTIONS

1. Fill up plastic bottle with water until it is almost full.

2. Add food coloring until it looks like a tasty fruit drink!

3. Now it looks just like a real soda or sports drink. Get your towel and pin ready.

4. Tighten the lid and place the bottle on the towel.

FAST FACT

There are lots of ways to color water. Food coloring is safe because it has been tested and approved by the Food and Drug Administration (FDA).

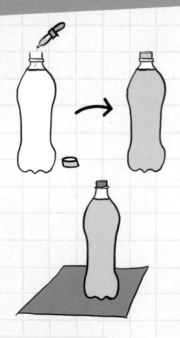

INSTRUCTIONS

5. Carefully hold the top of the bottle with one hand and, using the pin, poke six holes near the bottom of the bottle. (Some water might dribble out. Wipe it up with the towel.)

6. Once the water is wiped away, leave it out where your victim can see it.

✓ **Hold it by the cap!**

✗ **Don't grab the bottle here or the water with squirt out!**

HA HA HA!

🔍 DID YOU KNOW?

The shape that the water makes when it squirts out is called a parabola. All projectiles make a parabola when under the influence of gravity.

WHAT'S HAPPENING?

☑ For the water to start to move, there must be a new force when the lid is off.

☑ Air pressure (not gravity) gives the **extra** force that pushes the water out of the bottle.

DID YOU KNOW?

Water in your house is supplied by pressure produced at pumping stations. The earliest known water supply system was on the island of Crete in 2000 BC. Pressure in the water pipes was produced by gravity.

The amount of gravity acting on the bottle doesn't change. There has to be an increase in air pressure when the bottle is open!

Barrier

Force pushing back on water

Water trying to escape

Would this work in space where there is no air?

Extra force when the lid is open—the extra forces pushes the water out!

The same thing happens when the bottle is squeezed.

When the bottle is squeezed, the pressure on the water increases. Then, because it cannot be compressed, it is forced out of the holes.

FAST FACT

Weather can be determined by air pressure. High-pressure systems are sections of warm air that can hold more water. Low-pressure systems are sections of cold air that can't hold as much water—this leads to things like clouds and rain!

THE NEXT LEVEL: FOUNTAIN DEW

MORE SCIENCE!

The reverse of the Fountain Dew effect can also be true.

Put a bucket over your head and breathe normally.

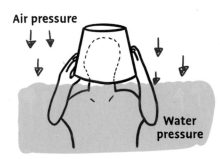

Air pressure

Water pressure

WATCH OUT!

The deeper in water you go, the more pressure there is. In the deep ocean, the air would get compressed by the stronger pressure of the water!

Lower your head under water. The bucket does not fill and you can still breathe. Phew! Thanks, physics!

Water pressure

Air pressure

(Air pressure and water pressure are balanced near the surface of water, so the water can't rush into the bucket.)

 FAST FACT

The feeling you get in your ears when they need to pop is known as ear barotrauma. It's caused by a difference in air pressure between your outer ear and middle ear. This puts pressure on your eardrum. Yawning or chewing gum can minimize the pressure difference by opening your eustachian tubes, which helps equalize the pressure in your ears!

DID YOU KNOW?

The physics of the diving bell was first described by Aristotle in the 400 BC, but not developed until the 1500s.

PRANK REVIEW: FOUNTAIN DEW

Be sure to document your reactions to Fountain Dew!

- What did you learn from the prank about air pressure?
- What did you learn from the prank about forces?
- When you're in an airplane, your ears might pop. Maybe that's why babies cry on takeoffs and landings. Where else can you see changes in air pressure?

EXPERIMENT #2

UPSIDE-DOWN GLASS

VICTIM: Sibling

FUNNINESS: 9

MESS: 8

SCIENCE: 10

DANGER: 4

SHOPPING LIST

- ☐ Clear glass with smooth rim
- ☐ Coaster
- ☐ Towel
- ☐ Water

WARNINGS TO FUTURE ME

- Practice flipping the glass outside or near a sink with towels nearby!
- Be very careful with a glass cup. Wet glass is slippery!
- Help your victim clean up the mess! (It'll be a big one! HA HA!)
- Don't use a thick coaster. If you don't have a coaster, use anything thin and solid that covers the top of the glass and is allowed to get wet (recycled cardboard, plastic cutting board, etc.).

PRANK PROCEDURE:
UPSIDE-DOWN GLASS

INSTRUCTIONS

1. Fill up a glass with water.

2. Place the coaster over the top of the glass.

3. Press down firmly on the coaster with one hand. With the other hand, flip the glass over and gently place it on the table or counter.

Pack extra towels or Mom will get mad!

NOTE
Make sure to practice this near a sink a few times first!

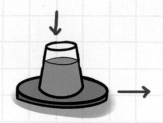

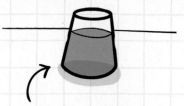

The water should stay in the glass without leaking! (Newton's First Law?)

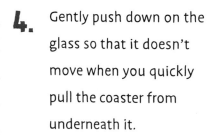
4. Gently push down on the glass so that it doesn't move when you quickly pull the coaster from underneath it.

5. There's only one thing left to do...wait for someone to clean it up! HA!

DID YOU KNOW?

Isaac Newton was a seventeenth-century scientist whose equations are still used when sending satellites into orbit around the Earth and probes to other planets.

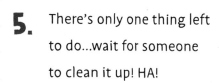

FAST FACT

Newton's First Law is referred to as the Law of Inertia. It states that an object at rest or in a constant state of motion will remain that way unless acted upon by another force.

WHAT'S HAPPENING?

Putting the coaster on the glass traps the water inside.

Upside down, the weight of the glass and air pressure keep the water inside.

The coaster can be moved without moving the water and glass because of Newton's first law of motion. Pull the coaster fast enough and there is no force pulling the glass.

When the glass is lifted, gravity pulls the water down and beats the air pressure trying to keep it up.

NOTE

May not work if the table surface is too rough or is badly chipped!

THE NEXT LEVEL: UPSIDE-DOWN GLASS

MORE SCIENCE!

You can do a similar experiment with the same glass and a piece of paper. Perform the same steps, replacing the coaster with a piece of thick paper.

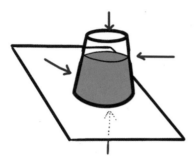

When you turn the glass upside down, the paper and water stay in place! Magic? No. Science!

The balance of air pressure between the inside of the glass and outside air keeps things in place. Air pressure pushes from all sides!

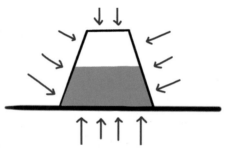

WATCH OUT!

This is the same effect that makes it difficult to lift the upside-down glass from the counter!

DID YOU KNOW?

Air pressure, or barometric pressure, is measured by a barometer. The oldest barometer was invented by the Italian physicist Evangelista Torricelli in 1643.

EVEN MORE SCIENCE!

A great example of air pressure is when we "evacuate" the air from a container with a pump.

A strong barrel can withstand the difference in air pressure.

However, if all the air is pumped out, it makes a vacuum.

Even steel barrels can be crushed by sucking the air out of them. That's how strong air pressure is!

DID YOU KNOW?

The first vacuum pumps were created in the 1600s and used to conduct scientific experiments. Today, vacuum pumps are used for many applications from televisions to fusion reactors!

The pressure at the bottom of the ocean is so large that when animals are brought up from its depths, they expand and sometimes have parts ooze out of them! Gross!

I am blob!

PRANK REVIEW: UPSIDE-DOWN GLASS

Be sure to document your reactions to Upside-Down Glass!

- What did I learn about Newton's First Law?
- What did I learn about gravity?
- If you turned the glass upside down at sea level and brought it to the top of Mount Everest (where the air pressure is less than one half of that at sea level), what do you think would happen?

EXPERIMENT #3

HUNGER EXPLOSION

VICTIM: Ketchup-loving family member

SCIENCE: 10

DIFFICULTY: 8

MESS: 9 (maybe 10!)

DANGER: 8

FUNNINESS: 8

SHOPPING LIST

- ☐ Ketchup bottle (about ¾ full), preferably a classic glass bottle for maximum effect
- ☐ Funnel or piece of paper
- ☐ Water
- ☐ Baking soda
- ☐ Vinegar
- ☐ Towel

WARNINGS TO FUTURE ME

- This can get very messy! Clean up fresh messes right away!
- Work quickly! Put the lid on straight after mixing the ingredients.
- Test the reaction with vinegar and baking soda.
- Don't add too much baking soda to the ketchup bottle or it won't burst!

BAKING SODA

PRANK PROCEDURE:
HUNGER EXPLOSION

INSTRUCTIONS

1. Open the ketchup bottle (make sure it is ¾ full).

2. Use a funnel or rolled up paper (to limit the mess).

3. Add just a little bit of water.

4. Add two spoonfuls of baking soda. Close the lid quickly!!

DID YOU KNOW?

The word ketchup comes from the Chinese word *kê-tsiap*. There are many "ketchups" but tomato ketchup is thought to be invented by American scientist James Mease in 1812.

NOTE

This works best when there is food. Order some French fries!

INSTRUCTIONS

4. With the lid on, shake the bottle.

5. When bottle is opened, the ketchup should explode out of the bottle!

HA HA HA!

HOW THE FINAL PRANK SETUP SHOULD LOOK:
yummy hot fries, shaken ketchup bottle, small empty bowl for the ketchup to be poured into.

FAST FACT

We measure how acidic or basic something is by measuring its pH. The pH scale usually goes from 0 to 14, with 0 being the most acidic and 14 being the most basic. Something like ketchup is acidic. This means that it will have a pH closer to 0. Baking soda is considered basic, so it will have a pH closer to 14. Something with a pH of 7 is neutral.

SOME NOTES

☑ Don't eat the ketchup afterward... It will taste yucky!

☑ Replace the ketchup bottle with a new one.

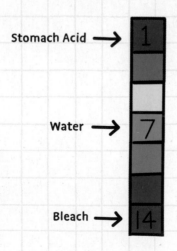

Stomach Acid → 1

Water → 7

Bleach → 14

SUPER ACIDS!

Certain acids can actually have a pH less than zero. For example, fluoroantimonic acid (the strongest known acid in the world) has a pH of -18.

WHAT'S HAPPENING?

The ketchup and baking soda react, just like the vinegar and baking soda. Why? Ketchup has vinegar in it!

Ketchup is an acid, and baking soda (sodium bicarbonate) is a base. Acids and bases often react to "cancel" each other out.

This experiment is the same as the famous science-fair volcano, which uses the baking soda and vinegar you tested with.

(But ours is definitely better because it is a hilarious prank!)

DID YOU KNOW?

As of publication, the Guinness World Record for tallest baking soda and vinegar volcano belongs to the students of Elmfield Rudolf Steiner School (UK). Their 28-foot tall volcano yielded a 4-foot (1.32-meter) eruption using 200 liters of baking soda and vinegar!

sodium bicarbonate

Sodium bicarbonate
(baking soda)
+
acetic acid (vinegar)
=
water, CO_2
(carbon dioxide),
sodium acetate

acetic acid

water

CO_2

FAST FACT

Any carbonate will make CO_2 bubbles when mixed with any acid.

MORE SCIENCE!

The reaction that takes place during the prank is...

CO_2 bubbles

When the bottle is open, the carbon dioxide quickly escapes, and with it, the ketchup!

FAST FACT

Chemical reactions such as in this prank can produce gases, like carbon dioxide. Gases take up more volume than solids or liquids, which is why all the ketchup sprays everywhere as it tries to escape the bottle!

DID YOU KNOW?

The name "acid" has its roots in Latin and loosely translates to "sour." Acids generally have a sour taste (but we don't recommend eating all acids).

PRANK REVIEW: HUNGER EXPLOSION

Be sure to document your reactions to Hunger Explosion!

- What did I learn about acids?
- What did I learn about chemical reactions?
- What are some other examples of acid-base reactions?

EXPERIMENT #4

ANTI-COASTER

VICTIM: a "neat freak"

FUNNINESS: 10

SCIENCE: 10

MESS: 8

DANGER: 6

SHOPPING LIST

☐ Thick cork coaster

☐ Paper cup with bottom rim

☐ Strong, thin magnet (rare earth or neodymium magnet)

☐ Tape

WARNINGS TO FUTURE ME

- Strong magnets work best, but they are more dangerous!

- Have a towel ready to clean up the mess.

- Do it on a surface that is allowed to get wet!

(The anti-coaster!)

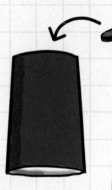

INSTRUCTIONS

1. If the magnet is not very thin, carve out a hole in the bottom of the coaster.

2. Tape the magnet down.

3. Flip over the coaster. (Can't see the magnet!)

4. Turn the cup over and tape the other magnet to the bottom.

ELECTROSTATICS

The tape sticks like two magnets but the force of attraction that makes tape work is not magnetic—it's electric! Electrostatic attraction keeps tape in place.

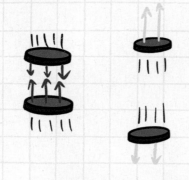

Correct!

Incorrect

HA HA HA!

INSTRUCTIONS

5. **THE MOST IMPORTANT THING**: Magnets attract OR repel. Make sure the cup's magnet is taped to repel the coaster's magnet.

6. Fill the cup with water and offer it to your victim.

7. When they try to place it on the coaster...

SOME NOTES

Some magnets look like this:

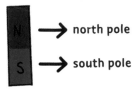 → north pole

→ south pole

Just like Earth!

Is the Earth a giant magnet?

Yes!

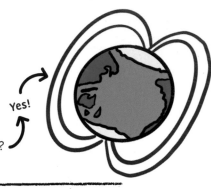

RULES FOR MAGNETS

↑

REPEL

↓

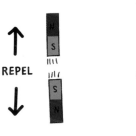

↓

ATTRACT

↑

All magnets have poles, even if they aren't painted blue and red!

N

S

The anti-coaster is a secret magnet.

And the cup is a secret magnet.

(Opposite direction!)

If the magnets were taped the wrong way, you'd have...

EARTH IS A MAGNET!

A compass works because it is a small magnet that aligns with the Earth, a big magnet!

The sticky coaster prank!

FAST FACT

Magnets stick to some metals by turning the metals into temporary magnets. (You can test this idea by lifting a magnet from a pile of paper clips.) These metals are called magnetic. The most common is iron—you can even find it in food! Don't worry it's perfectly safe to eat. (Test this too by seeing if you can attract your breakfast cereal to a magnet!)

MAGNETIC MAGIC!

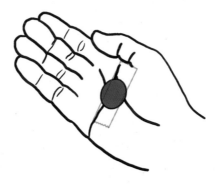

Tape a magnet on the palm of your hand. Place another magnet on a table. (Make sure the poles will face each other when you turn palm over!)

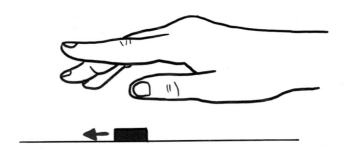

Wave your hand near the magnet... Are you a magician or a scientist?

FAST FACT

Like gravity, the magnetic force is invisible. Scientists use the idea of a *field* that exists everywhere in space to think about how forces act across distances. Magnets both change and react to the magnetic field.

James Clerk Maxwell (1831–1879) was a Scottish physicist who taught us that magnetism and electricity were two sides of the same force!

MASSIVE MAGNETS

The most powerful magnets in the universe are the collapsed cores of massive stars left behind after going supernova.

PRANK REVIEW: ANTI-COASTER

Be sure to document your reactions to Anti-Coaster!

- What did I learn about magnets?
- What did I learn about attractive versus repulsive forces?
- Can I use magnets to make something float in midair?

MENTOS MAYHEM

VICTIM: a soda stealer!

MESS: 8

DANGER: 5

FUNNINESS: 10

SCIENCE: 8

SHOPPING LIST

- ☐ Needle
- ☐ 4 inches floss
- ☐ Mentos (original flavor)
- ☐ 1 bottle Diet Coke with screw-top lid
- ☐ Scissors
- ☐ Towel

WARNINGS TO FUTURE ME

- Be careful when using a needle to thread the dental floss through the Mentos! Make sure an adult is supervising or helping.
- Keep the Mentos away from the cola. You can't reuse it if they touch!
- This is a guaranteed messy prank! Have towels nearby and always help your victim clean up. (After laughing, of course!)

roll of Mentos!

PRANK PROCEDURE:
MENTOS MAYHEM

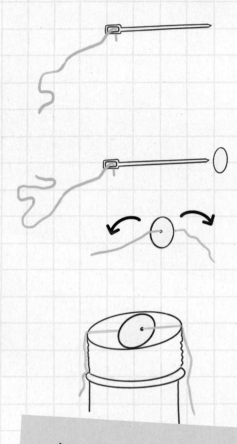

INSTRUCTIONS

1. Thread a needle with the floss.

2. Thread the floss through a piece of Mentos candy. (Ensure you have an equal amount of floss on each side of the candy.)

3. Twist off the cap of the soda bottle and stretch the floss across the open mouth of the bottle so the candy is held in place over the opening, and hold tight the ends of floss hanging down the sides of the bottle.

DID YOU KNOW?

Coca-Cola was invented in 1886 by American pharmacist John S. Pemberton (1831–1888) as a tonic made from coca leaf and kola nut.

INSTRUCTIONS

4. Place the lid back on the bottle and tighten. Cut off the overhanging floss so your victim cannot see it.

Inside the bottle

5. Place the cola in its usual spot (the fridge perhaps) and wait...

6. When your victim opens the bottle, the Mentos will fall into the cola and...MAYHEM!

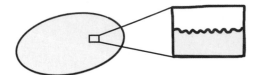

WHY MENTOS AND DIET COKE SPECIFICALLY?

Mentos appear smooth but are actually very rough under a microscope—they're porous!

Sodas are carbonated drinks. They have carbon dioxide dissolved in them.

The carbon dioxide reacts with the Mentos by fitting into all the tiny pockets.

The fizz you see and hear when opening a bottle is the carbon dioxide escaping!

Diet Coke works best because it has additives that make it fizzier!

DID YOU KNOW?

The first carbonated drink was invented by Dr. Joseph Priestley in 1767. He would spend many of his days inhaling unknown gases and then recording how they made him feel. One day, one of the gases made him laugh... he had discovered nitrous oxide, aka laughing gas.

WHAT'S HAPPENING?

The reaction between Mentos and Diet Coke is a physical reaction.

TWO TYPES OF REACTION:

Physical: when particles are moved around, but remain unchanged.

Chemical: when the particles are broken up and remade into completely different particles.

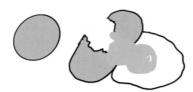

An egg breaking is a **physical** change.

Burning wood is a **chemical** change.

Chemical reactions change the chemical formulae of the reactants.

WOOD + OXYGEN $\longrightarrow$ CO_2 (carbon dioxide)

+

H_2O (water)

+

Heat ("exothermic")

DID YOU KNOW?

Mentos were invented in Poland in 1932. The shape of the candy is an oblate spheroid.

MORE SCIENCE!

Carbon dioxide was inside your drink and released when you opened it. Carbon dioxide is a chemical we find everywhere. In fact, your breath contains carbon dioxide made by your body!

O_2

CO_2

Carbon dioxide is called a "greenhouse gas" because it traps heat like the glass of a greenhouse, or a big blanket. Without it, Earth's atmosphere would be too cold. But with too much, Earth's atmosphere (and everything in it) will get too warm.

So we must limit things that create lots of carbon dioxide, such as burning fuel for transportation and electricity.

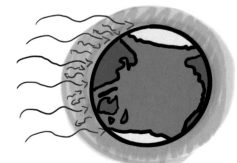

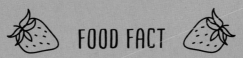

FOOD FACT

Diet Coke has no sugar, but is very sweet due to the use of artificial sweeteners. Contrary to popular belief, however, sugar content is not what makes food healthy or unhealthy. For example, many fruits are high in sugar but are still healthy for you.

DID YOU KNOW?

The "soda geyser" is a famous science demonstration for TV variety shows. It was first shown on the *Late Show with David Letterman* in 1999.

PRANK REVIEW: MENTOS MAYHEM

Be sure to document your reactions to Mentos Mayhem!

- What did I learn about physical changes?
- What did I learn about chemical changes?
- What would happen if the soda was left open overnight before trying the prank?

EXPERIMENT #6

EDIBLE POOP?

VICTIM: parent or adult

MESS: 7

DANGER: 10

FUNNINESS: 9

SCIENCE: 3

SHOPPING LIST

- ☐ 1½ cups powdered sugar
- ☐ ½ cup peanut butter
- ☐ ⅓ cup cocoa powder
- ☐ 1 tablespoon milk
- ☐ Crushed nuts
- ☐ Large mixing bowl
- ☐ Large mixing spoon

WARNINGS TO FUTURE ME

- Always check with an adult before using dairy or nuts in a prank. Your victim might be allergic!
- Place the fake poop where only your victim will find it. Don't put it in public or somewhere young children might find it.
- If you don't handle food properly by cleaning utensils, washing your hands, and placing the fake poop on clean surfaces, you could make yourself and your victim ill. And don't let the poop sit there for too long before eating it, or forget about it altogether and leave it there forever! That's how you get ants.

PRANK PROCEDURE:
EDIBLE POOP?

INSTRUCTIONS

1. Throw all the ingredients into a large bowl. (You can still perform the prank without the nuts. Ask an adult.)

2. Stir with a large spoon for about two minutes. (It should be dark brown with no lumps.)

SAFETY! Wash your hands before doing this step

3. Once the mix seems about the thickness of Play-Doh, you can start to mold it. (If not, add small amounts of milk or cocoa powder until it is the right thickness.)

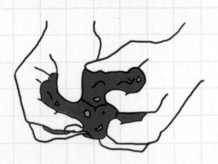

INSTRUCTIONS

4. Mold your fake poop into a realistic poop shape. (Don't pretend like you haven't seen poop before!)

5. "Hide" the poop in a clever location.

6. Let them find the poop.

7. You take a bite of it!

DID YOU KNOW?

You can change the color of your (actual) poop by changing your diet or eating certain foods. Poop is usually brown because it contains bits of old, used-up red blood cells.

THE PERFECT POOP DESIGN

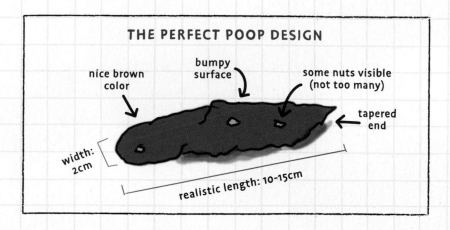

nice brown color

bumpy surface

some nuts visible (not too many)

tapered end

width: 2cm

realistic length: 10-15cm

SUGGESTED LOCATIONS:

- ☑ Kitchen floor
- ☑ Living room floor
- ☑ Kitchen counter
- ☑ A chair

(Remember to place the poop in a **clean** spot! Don't eat food that has been sitting out either. Also, don't eat real poop.)

WHAT'S THE DEAL WITH POOP?

Your poop is often referred to as excrement. Poop is discharged from your bowels. It's the stuff your body doesn't need.

There are billions of 'microbes' in your gut that your body needs to help it break down food. (That's good!)

Your body contains more microbial cells than it does human cells. But not all microbes are good. Some can make us sick, like E. coli.

E. coli is found in poop. But it will only hurt you if you get it in your mouth.

DID YOU KNOW?

In the 1880s, germ theory replaced miasma theory, which held that "bad air" caused diseases. We now know diseases are caused by germs like bacteria and viruses.

🔍 DID YOU KNOW?

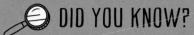

The microbial world of bacteria and protists was discovered in 1676 by Antonie van Leeuwenhoek who named them "animalcules" or "little animals."

There are more microbes that can make us sick too!

Salmonella   **Giardia**   Cryptosporidium

You can help prevent getting yourself and others sick by washing your hands before and after going to the bathroom and before eating.

Cooling or freezing food helps slow down the rate at which microbes multiply, which means "make copies of themselves."

THE NEXT LEVEL: EDIBLE POOP?

MORE SCIENCE!

- **MICROBIOLOGY** is the study of microscopic organisms. (Microscopic means they cannot be seen with the naked eye.)

- Microbiologists can look at how small microorganisms behave to better understand our own genetics and the spread of disease.

- The study of the spread of disease is known as epidemiology.

- Microorganisms are responsible for millions of preventable deaths each year. (Especially in developing countries where there is little access to clean water.)

DID YOU KNOW?

Thomas Crapper is often credited with inventing the flushing toilet. This is not true, but he did invent the tank filling float valve mechanism, which is still used in toilets today.

POOP SCIENCE

The average human poop has around one hundred billion bacteria per gram. Bacteria and viruses make up between 24 and 56 percent of your poop depending on the individual.

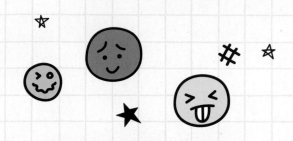

PRANK REVIEW: EDIBLE POOP?

Be sure to document your reactions to Edible Poop?!

- What did I learn from this prank about poop?
- What did I learn from this prank about hygiene?
- What part of the food that you eat makes your poop hard or soft?
- Whose response to the poop was the most surprising?

EAZY SQUEEZY

VICTIM: a very thirsty person

FUNNINESS: 8

SCIENCE: 7

MESS: 5

DANGER: 5

SHOPPING LIST

☐ Plastic water bottle with wrap-around label

☐ Strong rubber bands

☐ Towel

WARNINGS TO FUTURE ME

- A dry summer day in the park is the perfect place for this prank!

  (It's just water, so once it dries, all is well!)

- As always, if a mess is made, help clean it up.

- Replace your thirsty victim's water bottle with a new one.

  (One that is not pranked!)

- Be prepared for retaliation! (That's what the towel is for.)

PRANK PROCEDURE:
EAZY SQUEEZY

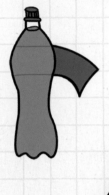

INSTRUCTIONS

1. Carefully remove the label from the bottle. You will need to place it back on.

2. Place as many rubber bands as you can around the bottle where the label was. Keep them as straight as possible!

3. Carefully put the label back on. (The rubber bands should be hidden.)

The Sun is actually white but I can't draw that on a white page!

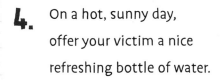

INSTRUCTIONS

4. On a hot, sunny day, offer your victim a nice refreshing bottle of water.

5. When your victim opens the bottle, the water will gush out! Ha ha!

HA HA HA!

QUICK QUIZ
Pressure increases with force and decreases with area. Bigger area, less pressure. Smaller area, more pressure. Do you think a bottle with a large spout will gush more or less?

EAZY SQUEEZY   53

RUBBER BAND SCIENCE!

Elastic material. It "wants" to be this size.

If it is stretched...

a force brings it back to its original size.

Lid ON vs. Lid OFF

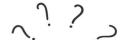

Why does it squeeze
the bottle only when
the lid comes off?

DID YOU KNOW?

The natural rubber band was patented by Stephen
Perry in 1845 to hold papers and envelopes together.

→ PRESSURE! ←

Water cannot be compressed. (But air can!)

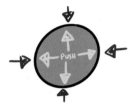

CAP ON

- Squeeze in from rubber band = push out from water pressure
- Bottle keeps shape

CAP OFF

- Squeeze in from rubber band = push out from water pressure
- Bottle loses shape
- The volume of the bottle shrinks, but the volume of the water stays the same, so the water must come out of the bottle!

FAST FACT

Most rubber bands are made from a liquid that comes out of a special type of tree (the rubber tree).

THE NEXT LEVEL: EAZY SQUEEZY

TEST!

Question: What will adding heat do to a rubber band?

Hypothesis: stretch/shrink/nothing

Need: Wall hook, hanger (with light clothing on it), pencil

1. Put rubber band in freezer (2 minutes)

2. Hang it on hook with hanger

3. Mark how much it stretched

4. Wait for it to warm up

5. Mark again

What happened?

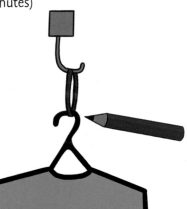

FAST FACT

The study of heat and how it can be used to do work is called thermodynamics and has been studied by scientists and engineers since harnessing the power of fire!

DID YOU KNOW?

Indigenous people of South America were the first people to use rubber. They made rubber balls to play a game known simply as Ballgame which could get quite violent and bloody!

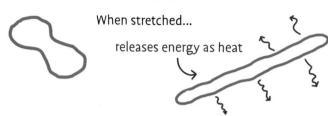

MORE SCIENCE!

When stretched...

releases energy as heat

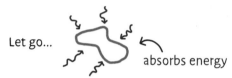

Stretch the band while it touches your lip to prove it!

Let go...

absorbs energy

PRANK REVIEW: EAZY SQUEEZY

Be sure to document your reactions to Eazy Squeezy!

- What did I learn from this prank about forces?
- What did I learn from this prank about energy and heat?
- What happened to the rubber band in the experiment and why?

EXPERIMENT #8

THE WET ONE

VICTIM: best friend

FUNNINESS: 11

MESS: 7

SCIENCE: 5

DANGER: 5

SHOPPING LIST

- ☐ Ten pencils (sharpened!)
- ☐ Strong clear plastic bag (sandwich bag)
- ☐ Thin plastic bag (vegetable bag from grocery store)
- ☐ Towel

WARNINGS TO FUTURE ME

- Sharp pencils are...sharp!
- Choose age-appropriate victim.
- Poke the bags carefully.
- Use your own pencils and bags. Don't ruin someone else's stuff. People love their stuff.
- Clean up the water for your victim! (It's not their fault you're the science genius of the two!)

PRANK PROCEDURE:
THE WET ONE

strong plastic (for you)

weak plastic (victim)

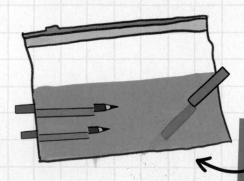

INSTRUCTIONS

1. Fill both bags half full with water.

2. Keep the strong plastic bag for yourself and give the flimsy plastic bag to your victim.

3. Poke several pencils through your bag (leaving the pencils in the bag).

4. The strong plastic bag will NOT leak!

SCIENCE FACT

The pencils will look broken because of the refraction of light as it passes from water to air.

INSTRUCTIONS

5. Tell your victim to do the same thing with their bag...

6. Water will leak everywhere! Ha ha!

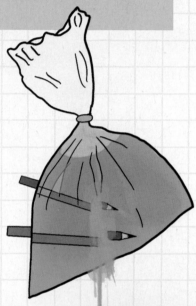

DID YOU KNOW?

Plastics can be natural or man-made. The first synthetic plastic ever made was called Bakelite and it was created in 1907.

Polyester sweaters from Patagonia are often made from recycled plastic bottles

PLASTICS 101

The numbers on the recycling symbol tell you what kind of plastic it is. Not all plastic can be recycled in one place, so check before you toss your plastic all in the same recycling bin.

There are lots of different kinds of plastics.

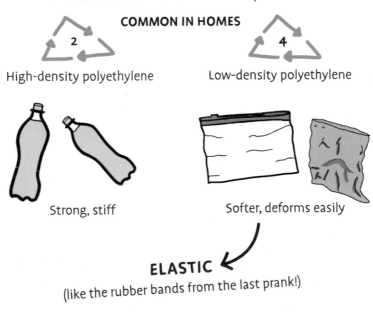

COMMON IN HOMES

2 — High-density polyethylene

4 — Low-density polyethylene

Strong, stiff

Softer, deforms easily

ELASTIC

(like the rubber bands from the last prank!)

Natural state

Stretched out

Snaps back

Long branching chains of molecules

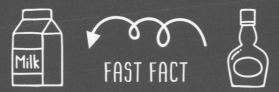

FAST FACT

You can extract a type of natural plastic from milk by mixing it with vinegar and filtering it out with a strainer.

DID YOU KNOW?

The word "plastic" comes from the Greek *plastikos* which means "to form." It was originally used to describe material that was easily shaped, but became a noun after the material became so widely used.

☑ Poke a hole in a sandwich bag and it tries to snap back.

☑ This makes a tight seal around the pencil so no water leaks. (Plastic is waterproof...duh!)

☑ The vegetable bag also stretches and snaps back, but it is weak. The long molecular chains break easily. The plastic is damaged and does not seal!

THE NEXT LEVEL: THE WET ONE

WHY IS PLASTIC A PROBLEM?

↓

It causes pollution to make plastic.

↓

Plastic ends up as pollution! It doesn't break down naturally, so it ends up in landfills or as litter.

REDUCE

REUSE

RECYCLE

WHAT CAN YOU DO?

REDUCE: buy less stuff, no "disposable" things (cups, utensils...), avoid overpackaged things (no need for extra wrapping on your apples!)

REUSE: donate toys/books to other families, use "trash" aka "found material" for art and crafts, use reusable bottles

RECYCLE: sort and recycle trash, and buy things that can be recycled or are made with recycled or recyclable materials

SAD FACT!

The Great Pacific Garbage Patch is a huge collection of tiny pieces of plastic caught in an ocean-sized vortex created by large ocean currents. No one yet knows just how much plastic is in our oceans.

The planet is yours to keep clean and healthy!

PRANK REVIEW: THE WET ONE

Be sure to document your reactions to The Wet One!

- What did I learn from this prank about plastic?
- What did I learn from this prank about recycling?
- What will I do to help reduce plastic use?

EXPERIMENT #9

AWARD-WINNING
SMILE

VICTIM: older brother

MESS: 8???

DANGER: 7

FUNNINESS: 9

SCIENCE: 6

SHOPPING LIST

☐ Food coloring
☐ Victim's toothbrush

WARNINGS TO FUTURE ME

- Don't spill the food coloring—it's really hard to clean out of stuff.
- Keep a spare toothbrush handy in case your victim really needs to clean their teeth.

HA HA HA!

PRANK PROCEDURE:
AWARD-WINNING SMILE

INSTRUCTIONS

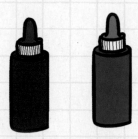

1. Choose your favorite color of food coloring or mix the colors (red would look scary, green would look gross)

2. Drop some food coloring onto the bristles of the toothbrush.

Too much and it will drip. Your victim will notice this! Too little and it will not change the color of your victim's teeth!

DID YOU KNOW?

Dr. John M. Harris founded the world's first school of dentistry in 1828. Today it is a museum in Bainbridge, Ohio.

INSTRUCTIONS

3. Remind your victim to brush their teeth. (Dentists recommend brushing at least twice a day for 2 minutes.)

4. The food coloring will look like a stain on your victim's teeth! (Don't worry! It's not permanent.)

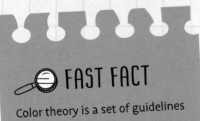

🔍 FAST FACT

Color theory is a set of guidelines for color mixing and description of the effects of mixing specific color combinations.

5. Don't forget to tell your victim how to remove the stain: brush teeth with a clean toothbrush and toothpaste!

WHAT ARE TEETH ANYWAY?

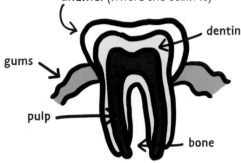

Enamel (where the stain is)

dentin

gums

pulp

bone

Enamel is porous (stains can get stuck in here)

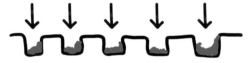

But food and food coloring don't stain
if you brush with toothpaste often!

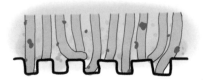

DID YOU KNOW?

Ancient Egyptians may have
started using a paste to clean their
teeth around 5000 BC, obviously
before the electric toothbrush
was invented (1927).

FOOD COLORING SCIENCE

TWO TYPES:

NATURAL: made from plants (example: sugar) and animals (example: beetles—yum!) Caramel color used in cola is made from sugars

ARTIFICIAL: made from human-made products (example: Red 40). Chemical formula: $C_{18}H_{14}N_2Na_2O_8S_2$. Used in cosmetics, candy, and drinks

WHY IS RED RED?

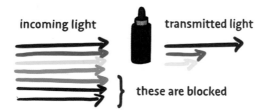

incoming light   transmitted light

these are blocked

COLOR FACT

White light is made of all colors, and all the colors combine to make white light. When you shine white light on red coloring, everything but the red light is absorbed, and the red is reflected back.

COLOR CHEMIST!

Try to create the color of cola with only artificial colors.

Need: red, blue, yellow food coloring, clear cups, sample cola, water

Start with a clean cup of water next to a cup of cola.

(Cola is the sample to compare to. Don't touch it.)

Slowly add different colors to try to match the cola

Results: Did it work? Can you guess if your color

is naturally or artificially colored?

COLOR FACT

Neuroscientists believe humans can perceive millions of distinct colors. They haven't all been named yet...so what are you waiting for?

FAST FACT

Green candy and ice creams are often colored using chlorophyll,
which is found in plants and gives them their green color!

QUICK QUIZ

In color theory, there
are primary, secondary,
and tertiary colors. Can
you name them all?

DID YOU KNOW?

Scientists have found evidence
for prehistoric humans using dyes
from over 35,000 years ago!

PRANK REVIEW: AWARD-WINNING SMILE

Be sure to document your reactions to Award-Winning Smile!

- What did I learn from this prank about natural food coloring?
- What did I learn from this prank about artificial food coloring?
- What difference does toothpaste make when brushing your teeth?
- Check the foods you eat and look for artificial coloring ingredients. How many are in things you ate today?

SECTION 2:
WANNA BET?

With great knowledge comes great power. Why not use your prank skills to "wager" your way to a week off your chores? Be warned, though: Sometimes it's not a good idea to bet for money, people get pretty upset when they find out they are being scammed, and scammers get pretty upset when they end up behind bars. Think outside the box! You could also try betting using other things, like speculating for a special treat or angling for more playtime.

In this section, you will find pranks that you can use to wager against unsuspecting victims. For these pranks to work well, you will have to be a bit of an actor as well as a mathematician. But even if you have trouble with your times tables and sometimes forget what 2+2 is, we have all the insider information to help you become the sleekest geek. Don't forget to fill out the "Prank Review." Pretty soon, you may never have to do chores again...

EXPERIMENT #10

CHEATER'S DICE

VICTIM: Math Teacher

FUNNINESS: 3

SCIENCE: 10

MESS: 0

DANGER: 3

SHOPPING LIST

- ☐ 4 wooden cubes
- ☐ 5 marker pens (4 colors + black)

WARNINGS TO FUTURE ME

- This is statistics, so it won't work every time, *so* you will need more rolls of the dice to increase your chance of success. This is how casinos make money!

- Don't take anyone's $$$... Not everyone can be a math genius like you.

FAST FACT

A "probability" of 0 means impossible, and a probability of 1 means certain to occur. All other chances of events are given a number between 0 and 1.

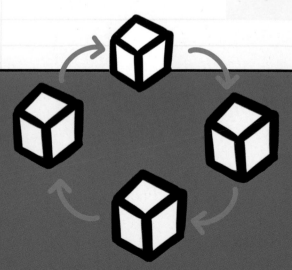

INSTRUCTIONS

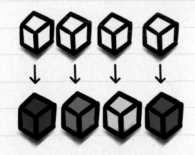

1. You will need to turn your four wooden cubes into six-sided dice, but first give them all different colors to easily tell them apart.

RED

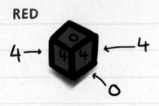

4→ 4 4 ←4
O
O

GREEN

2→ 2 2 ←2
6
6

2. You will then need to label them like this.

ORANGE

3 3
3 3 3
3

BLUE

5 1
5 5 ←1
1

3. Tell the victim that you will each choose a die and roll them. Whoever has a bigger number WINS!

FAST FACT

These are called non-transitive dice. Non-transitive dice can have many different number patterns. There are sets with fewer than four dice and sets with more. There are non-transitive twelve-sided dice too!

Green has the highest total value but...

Orange beats Green 2/3 times!

4. Allow the victim to look at each die. Then say, "You can pick first because it is an advantage."

It isn't! HA HA HA!

5. If they want to change colors, let them! You pick your die based on whatever one they choose according to the chart below.

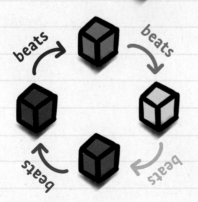

beats

beats

beats

beats

6. Roll the dice!

NOTES

In most games, there is a clear winner. Your victim will expect that the die with the biggest number on it should beat all other dice. But, the dice in this prank are a type of non-transitive dice called "Efron Dice." This means that every die has another die that beats it. I need to prove it!

RED sides: 0, 0, 4, 4, 4, 4

ORANGE sides: 3, 3, 3, 3, 3, 3

GREEN sides: 2, 2, 2, 2, 6, 6

BLUE sides: 1, 1, 1, 5, 5, 5

~~Math is hard!~~ Never mind, figured it out...

TRANSITIVE:

If x is related to y → And y is related z → Then x is also related to z

Example: 10 is more than 5, 5 is more than 1

So 10 must be more than 1 since numbers are **transitive**

The dice are NOT!!

QUICK STATS

The odds of any side showing up on a fair die is always 1 in 6, so even if you have rolled 3 a hundred times in a row, the odds of it being a 3 again are still 1 in 6.

DID YOU KNOW?

Dice have been found in many archaeological sites dating back to ancient civilizations. Egyptian hieroglyphs may have even depicted board games being played.

Red will beat **orange** 4 out of 6 times

If the victim chooses **orange**, always pick **red**.

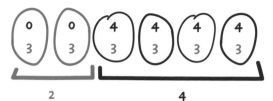

RED wins 4 out of 6 (2/3)

Orange will beat green 4 out of 6 times.

If the victim chooses green, always pick **orange**.

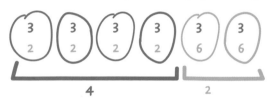

ORANGE wins 4 out of 6 (same as 2/3)

Green will beat **blue** 4 out of 6 times.

If the victim chooses **Blue**, always pick green.

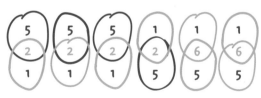

GREEN wins 8 out of 12 = 4/6

THE NEXT LEVEL: CHEATER'S DICE

So **blue** will beat **red** 4 out of 6 times?

Need to prove how...

5	5	5	1	1	1
0	0	4	4	4	4
1	1	1	5	5	5

Copy these numbers into your notebook
to prove **BLUE** wins!

FAST FACT

Paradoxes of probability
have shown how hard it
is for humans to grasp
numbers and chance.

QUICK QUIZ

The record for a die with the most sides is the disdyakis triacontahedron with 120 sides. That's 120 different possibilities. How many 6-sided dice would you need to match that number of possibilities? (Hint: it's much fewer than you might think!)

TRICKY DICE

Casinos try to make perfectly fair dice. Casinos don't
trick players, but they make bets that slightly favor the
house. Gamblers always lose in the long run.

DID YOU KNOW?

Florence Nightingale (1820–1910)
was famous for founding the
modern practice of nursing.
But she was also a statistician,
someone who analyzes and makes
conclusions from random data.

PRANK REVIEW: CHEATER'S DICE

Be sure to document your reactions to Cheater's Dice!

- What did I learn from this prank about probability
  and statistics?
- Which die did most people choose and why?
- What is a paradox and why are people so bad at
  seeing them?

EASY BUCK PICKUP

VICTIM: Uncle

FUNNINESS: 5

MESS: 1

SCIENCE: 7

DANGER: 3

SHOPPING LIST

- ☐ 2 cups (plastic for beginners)
- ☐ A grape (or Ping-Pong ball, or any small ball should work)
- ☐ A $1 bill (or just bet some candy)

WARNINGS TO FUTURE ME

- Use plastic cups instead of glass until you get good at this trick
- Using chores or candy is often better than betting money
- Make sure you don't hit the two cups together (especially if they're glass)
- Only use small, soft balls or else it could break the glass... Don't use a marble!

DID YOU KNOW?

You may hear the term "buck" when someone is referring to $1. Historians believe that the term "buck" traces back to when deer (buck) skin was once used as currency.

PRANK PROCEDURE:
EASY BUCK PICKUP

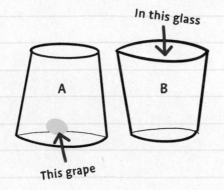

In this glass

A   B

This grape

1. Place a grape under one of the cups and bet your victims $1 that you can put the grape into the other cup without touching the grape or the second cup.

2. You can further encourage them to go along with the bet if you tell them you won't flip over the cup that has the grape under it.

3. Once your victim agrees to the bet, start to swirl the cup so the grape starts spinning around the inside edge. (It may take some practice to get the correct speed. You don't want it going too fast or too slow.)

GAME FACTS

This is called a "game of skill" as opposed to a "game of chance." Games of skill are ones where success is based mostly on physical or mental ability rather than luck.

INSTRUCTIONS

4. Continue swirling the cup at the same speed and lift it off the table and above the second cup

5. Stop swirling the cup and jolt it downward so the grape falls into the second cup. **Be careful not to bang the two cups together**.

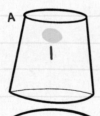

6. Take a bow and collect your buck.

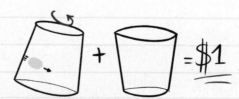

✏ WHAT'S HAPPENING? ✏

There is a force that keeps the grape going around
on the inside of the cup. It is the same force that you
can feel from the centrifuge ride at the fair.

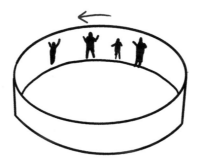

It may not look like it, but the grape is accelerating toward the center
of the cup! This is the same reason that Earth is held in orbit around
the Sun and that the Moon is held in orbit around the Earth!

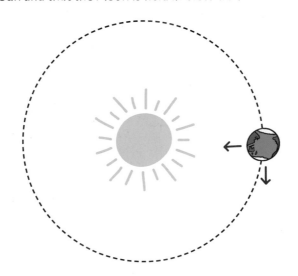

There are two forces working together that allow this trick to happen:

INERTIA: A property of matter that dictates how objects move. Inertia is affected by how heavy an object is and how fast it's moving. The bigger the mass, the harder it is to get moving, and also to slow down. This means heavy objects tend to have more inertia. Likewise, the faster an object is moving the more "inertia" it's said to have.

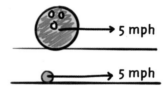

CENTRIPETAL FORCE: A force objects feel when moving in a curved path. Think tetherball. The grape is also feeling this force. Imagine looking at the grape from above.

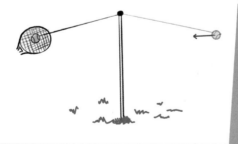

FAKE FORCES?

Centrifugal force is not real. It is just a sensation you feel because you're experiencing the centripetal force and inertial forces at the same time.

FAST FACT

Scientists may use a "centrifuge" to spin mixtures around in circles and separate out the different parts based on density.

MORE SCIENCE!

Inertia is the reason objects don't like to change direction.
The grape in the glass wants to travel in a straight line,
but the glass is forcing it to move in a circle.

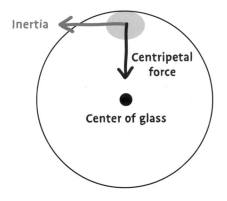

You also can do this with a (strong) bucket and water.
If you swing the bucket around with your arm, it
will not spill, even when it is upside down!

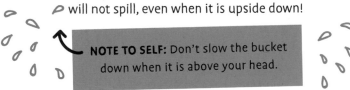

NOTE TO SELF: Don't slow the bucket
down when it is above your head.

The yo-yo trick "around the world" is a great example of
centripetal force. The yo-yo at the end of a string feels
centripetal force as it spins in a circle around your finger.

DID YOU KNOW?

Another place you can experience centripetal force is at an amusement park. Be thankful for the centripetal force as it prevents the roller coasters from flying off the tracks.

DID YOU KNOW?

Bakken in Denmark is the world's oldest amusement park. It has been continuously operating since 1583.

PRANK REVIEW: EASY BUCK PICKUP

Be sure to document your reactions to Easy Buck Pickup!

- What did I learn from this prank about inertia?
- What did I learn from this prank about centripetal force?
- Why don't race cars fly off the track when they speed around corners? What happens when they go too fast?
- What are some other examples of centripetal forces?

EXPERIMENT #12

MIND READER

VICTIM: Friend

FUNNINESS: 5

MESS: 0

SCIENCE: 10

DANGER: 0

SHOPPING LIST

☐ Only a calculator!

WARNINGS TO FUTURE ME

- Make sure the victim does the calculations correctly.
  - → Helps to speak SLOWLY and CLEARLY.
  - → Give them a calculator to be sure they are getting it right.
- When guessing the words, they might pick a strange one...this won't work all of the time.
- Set up the prank with confident showmanship... Touch your temple, close your eyes, make them believe you are reading their mind.

DID YOU KNOW?

Humans have been studying numbers and geometry since ancient times. Math and counting make up the earliest forms of writing!

Apple...?

Apple!

PRANK PROCEDURE:
MIND READER

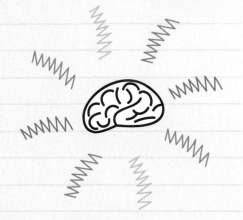

1. Tell your victim that you will be able to build an image by reading their mind, then say "I need to ask you some questions to get in sync with your brain waves."

2. **REMEMBER** this script:

Pick a number between 1 and 10.

Multiply it by 9.

Add the digits of that number together.

Subtract 5.

Think of the letter in that position of the alphabet.

Think of a country that starts with that same letter.

Think of an animal that starts with the last letter of the country.

Think of a color that begins with the last letter of the animal.

3. You should then say "There are no orange kangaroos in Denmark."

TROUBLESHOOTING

While the trick should work well, sometimes friends can be a bit slow or weird. Here are some tips for guiding them along.

☑ If they get the wrong letter for the country, they have done their math wrong. Give them a calculator.

☑ If they pick a weird color like "orchard pink" instead of orange, it will only partly ruin the trick. Go faster so they don't have too much time to think of weird responses.

☑ Victim may not know the country Denmark, or other clues; you can plant ideas in their head by dropping hidden clues days before trying the prank.

"Did you know the Danish pastry isn't actually from Denmark? (It's Austrian.)"

"The Danish flag is the oldest flag in the world!"

Denmark!

"Kangaroos can jump over twenty feet in a single leap!"

MATH + PSYCHOLOGY = MAGIC!

This prank works because it doesn't matter what number the victim picks, they will always have the letter D.

For example, say they pick the number 3.

Multiply by 9: 3 x 9 = 27

Add digits together: 2 + 7 = 9

Subtract 5: 9 − 5 = ④

1 x 9	=	9	→	0 + 9	=	9	→	9 - 5	=	4
2 x 9	=	18	→	1 + 8	=	9	→	9 - 5	=	4
3 x 9	=	27	→	2 + 7	=	9	→	9 - 5	=	4
4 x 9	=	36	→	3 + 6	=	9	→	9 - 5	=	4
5 x 9	=	45	→	4 + 5	=	9	→	9 - 5	=	4
6 x 9	=	54	→	5 + 4	=	9	→	9 - 5	=	4
7 x 9	=	63	→	6 + 3	=	9	→	9 - 5	=	4
8 x 9	=	72	→	7 + 2	=	9	→	9 - 5	=	4
9 x 9	=	81	→	8 + 1	=	9	→	9 - 5	=	4
10 x 9	=	90	→	9 + 0	=	9	→	9 - 5	=	4

The fourth letter of the alphabet is D.

MATH FACT!

The study of the peculiar properties of numbers is called Number Theory and is only one branch of mathematics.

Countries that start with **D**:

- **Denmark** (most likely one)
- Djibouti
- Dominican Republic
- Dominica

Denmark ends in K.

Animals that begin with **K**:

- **Kangaroo** (most likely one)
- Koala
- Kookaburra
- Kakapo

Kangaroo ends in O.

Colors that start with **O**:

- **Orange** (most likely one)
- Orchard Pink
- Ocher (surely no one is picking this?!)

EXPERT TIP: If it looks like they're REALLY trying hard to think of a color, they've likely thought of a Koala or Kookaburra. You should say, "It's hard to think of a color that starts with an *a*, isn't it?"

Most people will pick **Denmark, Kangaroo, Orange**.

Needs to be believable, like a magician. Have to practice the "act."

MORE MATH MAGIC!

NEW SCRIPT:

- Pick a three-digit number. All of the numbers must be different (e.g., 123).

- Reverse the order of the numbers (e.g., 321).

- Now you will have two numbers. Subtract the smaller number from the larger number (e.g., 321 − 123 = 198)

- Add all the digits together (e.g., 1 + 9 + 8).

- You are thinking of 18.

HOW IT WORKS:

They pick number **abc** = 100**a** + 10**b** + **c**

abc − **cba** = 100**a** − 100**c** + 10**b** − 10**b** + **c** − **a**

$$= 100(\mathbf{a} - \mathbf{c}) + (\mathbf{c} - \mathbf{a})$$

$$= 100(\mathbf{a} - \mathbf{c}) - (\mathbf{a} - \mathbf{c})$$

$$= 99(\mathbf{a} - \mathbf{c})$$

There are only 9 possible solutions to this equation: 99, 198, 297, 396, 495, 594, 693, 792, 891…all of these add up to 18!

0 1 0 1 0 1 1 0 0 0 1 0 **FAST FACT** 1 0 1 1 1 0 0 1 0 1 0 1 0

Most people use the decimal counting system which includes 10 symbols for numbers, but there are other ways of counting too, including the binary system which only uses 1's and 0's, and even the duodecimal system, like the one we use to tell the time.

$$a^2 - b^2 = (a+b)(a-b)$$

MATH FACT!

Replacing numbers with symbols that can stand in for any number is the topic of algebra. These symbols are called variables. It can be as easy as A + B!

DID YOU KNOW?

The term "algebra" is derived from *al-jabr*, which comes from a book by the Persian mathematician Muhammad al-Khwarizmi (780–850). His name also is where the term "algorithm" comes from.

PRANK REVIEW: MIND READER

Be sure to document your reactions to Mind Reader!

- What did I learn from this prank about multiples of 9?

- What did I learn from this prank about number theory?

- What other patterns can I find in numbers?

- The solution to 11 x any two-digit number can be found by splitting the multiplied number then adding it together in the middle. For example, 11 x 23 = 253 because 2 + 3 = 5.

EXPERIMENT #13

YOUR COIN? MY COIN!

VICTIM: Friend

FUNNINESS: 7

MESS: 0

SCIENCE: 6

DANGER: 4

SHOPPING LIST

☐ 4 to 6 coins

WARNINGS TO FUTURE ME

- More exciting and surprising if your victim is super skeptical.
- Don't actually take anyone's money (pretend at first, then return it).
- Using props (like a blindfold) can make the trick extra convincing.
- Make sure the coins don't all start warm. (Don't be sitting on them!)

PRANK PROCEDURE:
YOUR COIN? MY COIN!

INSTRUCTIONS

Victim

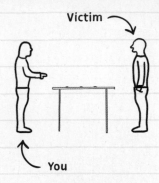

You

1. Sit across from the victim with your coins on the table.

2. Tell them you will turn around, and then ask them to choose ONE coin by picking it up.

"If I can correctly pick which coin..."

3. Say "If I can correctly guess which coin you picked up, I get to keep all of the coins."

4. Tell them to put their coin back on the table once they are sure which one it is. To make it more believable, you can tell them to shuffle them if they want.

☼ TIP ☼
Keep talking so they hold it for longer. You can tell your victim that it's important they transfer their essence into the coin so you can sense it.

TIP

Words like "resonance" and "magnetic attraction" help to sell it.

5. Tell them "I will now use my magical powers to find your coin."

6. Touch each coin GENTLY with your finger as you talk about your magic powers, but don't take too long!

7. One of the coins will be warmer than the others. This is your victim's coin.

8. Pick up the warm coin and present it to them. "Is this your coin?"

"Is this your coin?"

INVITE THEM TO GIVE IT A GO

☑ Hold the coin lightly with only your fingertips to not warm it up (if your victim is clever).

☑ Next time, warm up the coin to see of your victim can figure it out. (So generous!)

WHAT HAPPENED?

Everything has atoms jiggling and moving around inside it.

HAND

Atoms move fast

More energy

Higher temperature

COIN

Atoms move slow

Less energy

Lower temperature

Faster atoms in the hand hit the slower atoms in the coins

Coin heats up

Place on hand where coin is cools down

SECOND LAW OF THERMODYNAMICS

Heat travels from something hot to something cold.

Heat CANNOT travel from something cold to something hot.

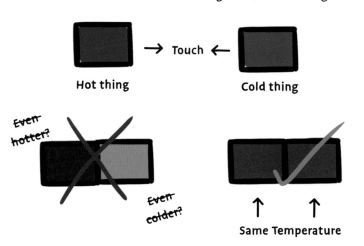

Heat energy moves from the hand to the coin --> Atoms in the coin move faster --> More energy in the coin means higher temperature.

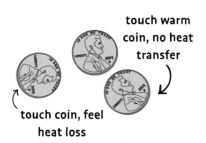

touch warm coin, no heat transfer

touch coin, feel heat loss

FAST FACT

A piece of metal will feel colder than a piece of plastic or wood at the same temperature because metals are better conductors of heat, which means they are better at taking heat away from your body.

MORE THERMODYNAMIC MAGIC!

CRUSH A BOTTLE WITH ~~YOUR MIND~~ PHYSICS!

1. Have an empty plastic bottle with the lid on.

2. Announce "I will now crush this bottle with the power of my thoughts!"

3. Place in freezer/snow in winter/ice bucket and (pretend to) think really hard.

4. The bottle will COLLAPSE.

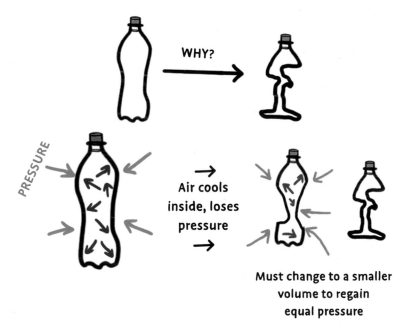

WHY?

PRESSURE

→ Air cools inside, loses pressure →

Must change to a smaller volume to regain equal pressure

HOT TIP!

When metals cool down, they take up less space (contract) and when they heat up, they take up more space (expand). Same with gases. The opposite is true of rubber bands! (Remember the Eazy Squeezy prank?)

DID YOU KNOW?

Ludwig Boltzmann (1844–1906) was an Austrian physicist who developed statistical mechanics, which tells us how the movement of microscopic atoms and molecules give rise to temperature, volume, and pressure. His famous equations are even inscribed on his tombstone.

PRANK REVIEW: YOUR COIN? MY COIN!

Be sure to document your reactions to Your Coin? My Coin!

- What did I learn from this prank about thermodynamics?
- What did I learn from this prank about metal as a heat conductor?
- Does the size of the coin affect how much it heats up?

EXPERIMENT #14

MOON VISION

VICTIM: Friend

FUNNINESS: 5

MESS: 0

SCIENCE: 7

DANGER: 2

SHOPPING LIST

- ☐ Full moon (in the sky)
- ☐ Camera
- ☐ Ruler

WARNINGS TO FUTURE ME

- Do not try this illusion with the Sun. It will damage your eyes!

- Be careful where you view the Moon from. Don't stand in the middle of the road!

- You can use a moon chart to check when the next full Moon will be.

- Check the weather—this won't work if it's cloudy and you can't see the Moon!

PRANK PROCEDURE:
MOON VISION

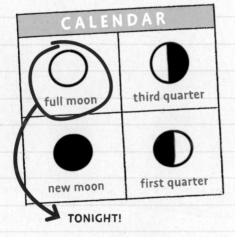

CALENDAR

full moon | third quarter

new moon | first quarter

TONIGHT!

1. Use a lunar calendar to figure out when the next full moon will be.

2. **Early in the evening** (just after sunset) take a friend outside and WOW them with facts about the Moon:

   → **"The same face of the Moon is always pointing toward the Earth."**

   → **"The Moon causes the Earth's tides."**

   → **"The Moon is slowly drifting away from the Earth."**

3. Make a comment on how big the Moon looks, then set up your victim by asking them, "When do you think the Moon is bigger? Early or later in the evening?"

Katherine Johnson was an American mathematician who helped break down racial and gender stereotypes and help NASA land the first people on the Moon.

4. Respond with something like, "I think it might be the same size."

5. Use a ruler to measure the Moon by holding your arms straight out and taking a photo of the ruler over the Moon.

6. Set an alarm for two or three hours later, when the Moon is higher in the sky. It will look smaller now, so if you want, now you can double the bet.

7. Use your ruler and outstretched arms to take another measurement and photo of the Moon, just like the first one. This is known as a control.

➡ WHAT'S HAPPENING? ⬅

The Moon always appears larger when it is close to the horizon. Scientists today still aren't 100 percent sure why this illusion occurs.

If your victim still doesn't believe you (or to check it out yourself), when the next full moon occurs and it is closest to the horizon, have them look at it between their legs. No, this is not a prank.

This will work the best when the Moon is nearest to the horizon.

The Moon should look smaller!

SCIENCE TIP!

When scientists use instruments (like rulers) to take reading using numbers, it is called a "quantitative" measurement.

FAST FACT

Often we can't trust our eyes or our senses, so scientists use specially designed equipment to take measurements... This makes science more reliable.

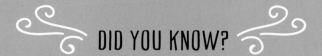

DID YOU KNOW?

One of the oldest and most famous optical illusions appeared in *Harper's Weekly* in 1892 called "rabbit duck."

MORE SCIENCE!

Things always appear bigger when there are other things nearby that you can compare them to (like the trees and buildings near the horizon).

You can use the image below to bet a victim which black circle is larger!

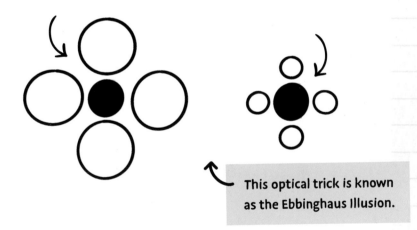

This optical trick is known as the Ebbinghaus Illusion.

Both circles are actually the same size!
This shows how the location of an object can
influence our perception of its size!

Below is a picture of a road with two orange shapes inside it. Which orange shape is longer?

This trick is known as the Ponzo Illusion.

Surprise! They are the same size again! Don't believe it? Use a ruler.

While these two illusions don't fully explain the Moon Illusion, they are great examples of how easily our perceptions can be fooled!

QUICK QUIZ!

There are not just optical illusions, but also auditory illusions. Can you guess which of your senses are being tricked with an auditory illusion and why one doesn't appear in this book?

DID YOU KNOW?

Galileo Galilei is credited with outlining the various phases of the Moon in 1609. His watercolor drawings are considered the first realistic depictions of the Moon.

FAST FACT

Lunar eclipses only happen during full moons because the Sun must be on the opposite side of the Earth for them to occur.

PRANK REVIEW: MOON VISION

Be sure to document your reactions to Moon Vision!

- What did I learn from this prank about optical illusions?
- What did I learn from this prank about experimental controls?
- Why is it so important to use measuring tools, like rulers, to help observe the world?

SECTION 3:
CLEAN CLASSICS

These pranks are for the purist. They are simple, relatively innocent, and easy to clean up, as well as requiring few acting skills. These are the sort of pranks you can set up, sit back, and just watch unfold. The true skill of these pranks, however, lies in patience and planning.

You could need a day (or maybe a night) to set these pranks up. But once your creations are complete, there is nothing to do but ensure your victim finds them and uses them. You could test out how long it takes for them to figure it out, or just to lose their cool... Sometimes it might be best if you aren't in the room as their frustration grows.

EXPERIMENT #15

ALL-NATURAL SELECTION

 **VICTIM:** People at a party

 **FUNNINESS:** 8

SCIENCE: 10

 MESS: 2

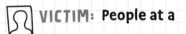 **DANGER:** 3

SHOPPING LIST

- ☐ Big bag of jelly beans (or other candies)
- ☐ Large serving bowl
- ☐ "Stop-bite" nail polish (available at any drug store)
- ☐ Pen and paper

WARNINGS TO FUTURE ME

- Check with an adult first, or else they may throw out the entire jelly bean bag.
- Wash your hands before handling any food.
- Allow the polish to dry before putting jelly beans back in the bowl, or else it may ruin all of them.
- Only paint ONE color.

PRANK PROCEDURE:
ALL-NATURAL SELECTION

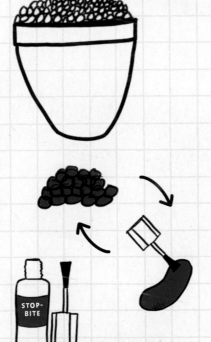

STOP-
BITE

INSTRUCTIONS

1. Wash your hands and pour all jelly beans into a clean serving bowl.

2. If you want, you can count how many of each color jelly bean there is in the bowl.

3. Take out all the red ones (or any color that everyone loves).

4. Paint all the red ones with the special "stop-bite" nail polish. You don't need to use a lot—just a dab on each jelly bean of the chosen color is enough.

DID YOU KNOW?

The gooey-chewiness of candies, like jelly beans for example, comes from pectin, gelatin, and starch mixtures. The technique was first developed in Germany in the early 1900s at the Haribo company.

INSTRUCTIONS

5. Let the polish dry before mixing them back into the bowl, then wait...

6. Word will quickly spread that there is something wrong with the red jelly beans.

7. At the end of the party, you can see how many of each type of jelly bean remains.

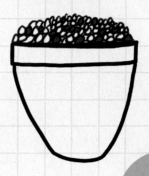

I heard they taste awful!

Yuck! Don't try the red ones!

If you keep doing it, word will spread further, and no one will *ever* eat the red ones again—they will always be all yours!

WHAT'S HAPPENING?

A small change in how something tastes or acts can dramatically and quickly change how people behave around that thing.

A change in taste of a jelly bean only needs to be noticed by a couple people before no one wants to eat them anymore. People will tell others how bad they taste, and soon everyone will avoid them. Pretty soon, you can change your whole family's behavior to stop eating your favorite treat!

This is the same reason some insects are so colorful. They are warning predators that they taste bad or are poisonous.

FAST FACT

All the living things on Earth today make up only about 1 percent of all the life that has ever existed on our planet. That means 99 percent of all life-forms that once called Earth home are now extinct.

Over time, animals and plants have slowly changed to help them avoid being eaten.

- Some evolved camouflage to hide from predators.
- Some evolved strong and long legs to help them run away from predators.
- Some evolved spikes or spines to make it difficult to eat them.

Predators also evolved to make it easy to eat more stuff.

- Some evolved special vision to better see colors.
- Some evolved strong long legs to catch prey.
- Some evolved certain shaped teeth, beaks, and mouths to make it easier to eat some things.

Fast cheetah mom

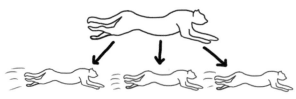

Fast cheetah cubs

The animals and plants with traits that helped them survive, passed these characteristics on to their babies. Those without special features did not survive or reproduce as successfully and died out.

DID YOU KNOW?

Life on Earth began around 3.8 billion years ago, and all the life-forms alive today are related.

 ## THE NEXT LEVEL: ALL-NATURAL SELECTION

This prank can be used on any kind of food. Try it on candy snakes, chocolate bars, pretzels... But make sure that whoever buys the food knows what's up or else they may just stop buying it altogether.

Natural selection has shaped how every living thing on the planet works and behaves with all of the other living and nonliving things on the planet.

⟩ FAST FACT ⟨

Lizards and birds have evolved to see colors to help them communicate and/or identify what is safe to eat and what is poisonous. Plants have also evolved to make their fruits colorful so animals will eat them and then poop the seeds out, which helps produce another plant!

DID YOU KNOW?

Evolution by natural selection was first proposed by Charles Darwin in 1859 to help explain why we see so many different types of creatures on Earth.

Little brother!

Me!

PRANK REVIEW: ALL-NATURAL SELECTION

Be sure to document your reactions to All-Natural Selection!

- What did I learn from this prank about natural selection?
- What did I learn from this prank about evolution?
- Why are some bugs colorful if they don't want to be eaten, but some fruits are colorful when they do want to be eaten?
- Can you think of some other examples of natural selection?

EXPERIMENT #16

SPILLED MILK

 **VICTIM:** Dad

 **FUNNINESS:** 8

 **MESS:** 5

 **SCIENCE:** 5

 **DANGER:** 4

SHOPPING LIST

- ☐ PVA glue (aka wood glue)
- ☐ Flat smooth surface (glass, tile, or marble work best)
- ☐ Soap or detergent
- ☐ Knife or scraper (get an adult to help)

WARNINGS TO FUTURE ME

- Ask an adult to help with the scraping because making the "spill" can get a bit sticky and messy.

- Be careful when handling the knife. It's not for cutting, it's just for scraping the glue off the glass.

- Be careful not to glue stuff together and always watch out when handling sharp glass or tiles.

PRANK PROCEDURE:
SPILLED MILK

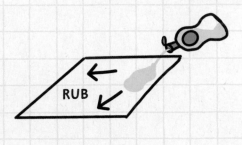

INSTRUCTIONS

1. Carefully lay flat the piece of glass and rub a thin layer of detergent over the surface.

2. Cover the surface of the glass with the PVA glue, making it look like a spill, then leave it so it can dry—probably a few hours. The thicker the layer, the easier it is to remove.

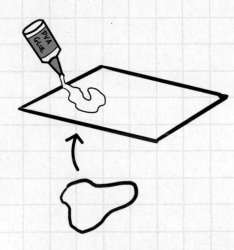

3. Later, test to see if the glue is dry by gently touching the middle of the spill with your finger. If you can see your fingerprints in it, it needs more time.

DID YOU KNOW?

The PVA in PVA Glue is polyvinyl acetate and was discovered in 1912 Germany by Fritz Klatte.

Add a toppled empty glass nearby to add to the effect.

INSTRUCTIONS

4. Once it is dry, use a blunt knife to gently lift the edges off the glass. Get an adult to help you with this step.

5. Place the "spill" on top of a computer, or anyplace it's a bad idea to have milk spilled.

SOME SCIENCE

People get worried about liquids (like milk and water) spilling on their electronic devices because they think it will ruin the electronics.

Water and milk actually contain impurities that are often charged particles. (Impurities aren't necessarily bad—sometimes they actually clean water and can help make your teeth stronger!)

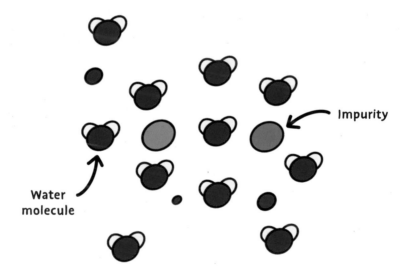

Impurity

Water molecule

Electronics work because of the charges in the wires, so adding charged particles can disrupt the charges in the device, causing short circuits and damaging it.

-ː MILK SCIENCE ː-

Your body contains an enzyme called rennin. Enzymes help chemical reactions happen faster than they would naturally occur. When you drink milk, rennin helps your stomach turn the milk into clumps for easier digestion. This is known as coagulation. A baby's stomach contains extra amounts of these enzymes because they drink so much more milk.

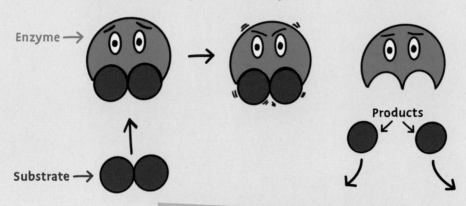

Enzyme →

Substrate →

Products

FART FACT

Milk contains a sugar called lactose that is different from other sugars. As we get older, we make fewer enzymes to digest lactose, making more gas in our guts.

Adults with lactase enzymes is a recent event in evolution thought to have occurred around 5,000 years ago. Which means there was probably a lot more farting going on.

Barry Marshall won a Nobel Prize in 2005. The research involved him drinking a broth of the bacteria *Helicobacter pylori* to prove that it was the cause of stomach ulcers.

THE NEXT LEVEL: SPILLED MILK

MORE SCIENCE!

You can make other types of spills using different glues.
You can even buy clear PVA. Just make sure the glue you
choose is safe to use and have an adult help you!

You can add food coloring and chunks to the glue to make fake vomit!

By putting the glass at a slight angle, you can give
the impression that the spill has run.

FAST FACT

Geckos are able to stick to surfaces
due to adhesive forces in their feet.
With their feet covered in millions of
microscopic hairs called "setae," geckos
have enough attractive force between
their feet and the surface they are
walking on to support their full weight.

Other kinds of spills could include:
- Juices
- Soft drinks
- Melted ice cream

Liquids have a fixed volume but can change their shape, which is why when they spill out of a cup, they go everywhere. Solids, on the other hand, have a fixed volume and fixed shape, so if they spill out of a cup they don't really do anything and can just be picked back up.

Where are some other places a spill would be found and be a problem?

PRANK REVIEW: SPILLED MILK

Be sure to document your reactions to Spilled Milk!

- What did I learn from this prank about the traits of wood glue?

- What did I learn from this prank about the conductivity of liquids?

- Why is it a bigger problem to spill stuff on some things?

- Is it okay to cry over spilled milk?

EXPERIMENT #17

SUDSLESS SOAP

 **VICTIM:** Mom or sister

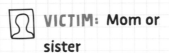 **MESS:** 2

DANGER: 5

FUNNINESS: 8

SCIENCE: 7

SHOPPING LIST

☐ Bar of soap
☐ Clear nail polish

WARNINGS TO FUTURE ME

- Do not get nail polish on your hands—it is almost impossible to get off! Wear gloves.
- Remove all the other soap from the room to ensure they use the prank soap.
- Put the nail polish away in a safe place where no one can get to it.
- Buy your own nail polish, because your mom will get mad if you use all of hers.

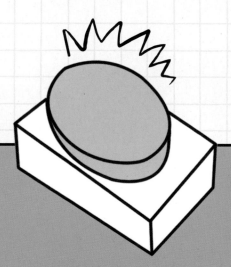

PRANK PROCEDURE:
SUDSLESS SOAP

INSTRUCTIONS

NOTE

This prank is pretty simple, but to ensure it works well, you need to be thorough with how you set it up.

1. Coat a brand-new, dry bar of soap with the clear nail polish. It must be entirely coated and should have a thick layer on it.

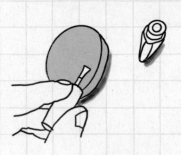

2. Let the nail polish dry and, if need be, apply a second coat. Remember, if any of the soap isn't covered, the prank won't work properly.

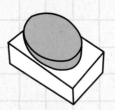

3. Remove all the other soap from the bathroom and place the prank soap in the shower. Don't forget to hide the nail polish.

QUICK QUIZ

There is a soap made in Lebanon made of pure gold dust and real diamonds. It costs $2,800 for one bar. Do you think it is better at cleaning your bum?

4. Wait for the victim to use the soap, but it's probably best not to be in the bathroom with them for this...

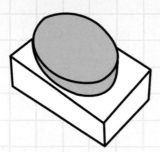

5. The soap should not be able to suds up, leaving the victim perplexed.

What the...?!

DID YOU KNOW?

Soap gets its name from "Sappo Hill" in ancient Rome where a stream flowed in which certain chemicals mixed allowing Roman women to clean their clothes more easily.

DID YOU KNOW?

Ancient Egyptians used soap to keep clean. But supposedly their soap helped to heal and soothe their skin. This is because Egyptian soap included olive oil as an ingredient, which moisturizes skin.

WHAT'S HAPPENING?

Oil and water are different types of molecules that can't interact with each other. This is why washing with just water doesn't clean very well.

SOAP MOLECULE

head →

← tail

Soap is made of a kind of special molecule made from a "head" and a "tail" that allows it to interact with both oil and water.

The "tails" of the soap molecules attach to the oils, which can now be washed away as the soap "heads" attach to the water.

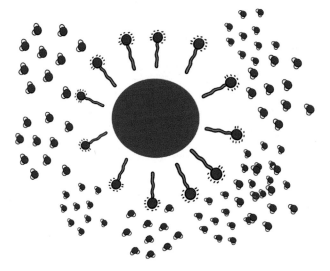

Soap gets "sudsy" because of how it interacts with water, air, and oils. The nail polish creates a barrier over the soap that prevents this from occurring.

The nail polish is made up of another kind of molecule that stays as a liquid until it is spread thin and allowed to dry. This molecule creates long, stiff chains that don't interact with water or oil.

DID YOU KNOW?

Soap operas were named because the advertising that supported the dramatic TV and radio series came from soap companies.

THE NEXT LEVEL: SUDSLESS SOAP

MORE SCIENCE!

The shape and structure of molecules is what gives certain materials (like soap, oil, and water) their properties. Wood is hard because of the specific molecules it's made of. Likewise, different plastics are rigid or soft depending on the type of molecule that they are made from.

Cell membrane

This area is the cell wall and it is quite different from the cell membrane.

The cells in our bodies have a special coating on the outside called a membrane that helps separate the stuff on the inside from the outside by acting a lot like soaps!

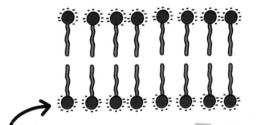

This formation is known as a *phospholipid bilayer*. When left in water all the molecules form a shape that is similar to the cells in your body!

FAST FACT

The average human adult body contains roughly 100 trillion cells! Some are packed tightly, like red blood cells, and some are spread thin, like those on your skin.

CLEAN FACT

We have to wash our hands to remove microscopic germs that love to live in the invisible oil layer on our hands, in the creases in our skin, and under our nails. Soap helps lift this oil layer off and allows it to be washed away by the water.

Soap doesn't "kill" germs on our hands like hand sanitizer does. However, it is much more effective at removing germs that live in natural oils made by special glands in our skin.

While it may sound better to "kill" these nasty germs, science teaches us that hand washing is the best way to stay healthy and stop the spread of disease-causing germs!

PRANK REVIEW: SUDSLESS SOAP

Be sure to document your reactions to Sudsless Soap!

- What did I learn from this prank about polar and nonpolar molecules?
- What did I learn from this prank about how soaps and detergents work?
- How does the size and shape of a molecule affect how a substance works?

REMOTE CONTROLLED

VICTIM: Brother

FUNNINESS: 7

MESS: 1

SCIENCE: 6

DANGER: 3

SHOPPING LIST

- ☐ Electrical tape or duct tape
- ☐ Scissors
- ☐ TV remote control

WARNINGS TO FUTURE ME

- The glue on the tape can leave behind a sticky mess—clean it up once the victim figures it out.
  - → Like a pro: Use rubbing alcohol (ask an adult for help)
  - → Newb: Use warm soapy water
- Make sure the tape is the same color as the TV remote (probably black)

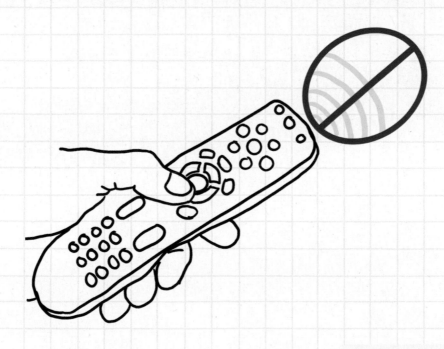

ACTUAL SIZE

INSTRUCTIONS

1. Cut a small square of tape. You may want to use a ruler to measure the size of the transmitter.

2. Find the transmitter on the remote control and cover it with the tape. If done correctly, the tape should be invisible.

TIPS AND TRICKS

The type of tape you use matters, you may need to do some trial runs with different types of tape first.

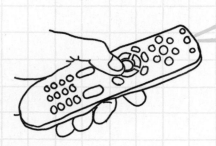

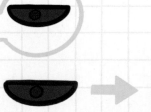

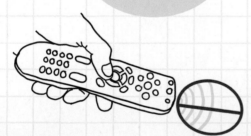

They may even try to change the batteries. HA HA HA!

INSTRUCTIONS

3. Put the remote back where you found it and wait...

4. When your victim tries to use it, it will not work and they'll think something is broken.

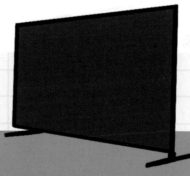

☀ DID YOU KNOW? ☀

The first televisions appeared in the 1920s and were based on cathode ray tubes (CRT). These TVs where analog, receiving continuous signals. Nowadays, TVs receive digital signals, which can be represented by long lists of 0's and 1's. Digital technology has many advantages, including the ability to have exact copies of information.

WHAT'S HAPPENING?

A remote control sends a signal to the TV using a special type of electromagnetic light called "infrared."

ELECTROMAGNETIC SPECTRUM

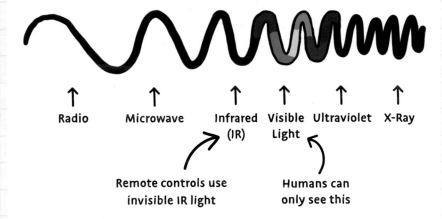

Radio   Microwave   Infrared (IR)   Visible Light   Ultraviolet   X-Ray

Remote controls use invisible IR light

Humans can only see this

⚡ FAST FACT ⚡

Satellites and telescopes look at the universe in different wavelengths, and we can see that different stars produce different types of electromagnetic energy.

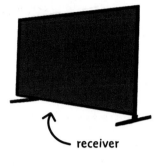

receiver

The remote can control the TV by shining light at it. Different pulses of light tell the TV to do different things, like turn the volume up or down.

Different tapes block different types of light. Not all tape will block infrared light either...clear tape doesn't work and it also doesn't block visible light... Duh...

NIGHT VISION

All animals give off infrared energy that we can sometimes feel as heat. Night-vision goggles use infrared energy to allow their wearers to see in the dark.

Electrical tape or duct tape won't block cell phones or Wi-Fi because they use radio waves, which can even go through walls!

MORE SCIENCE!

Most computer mice are "optical."

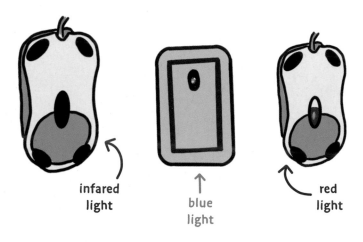

infared
light

blue
light

red
light

Some use infrared light, some use blue light, some use red light.

The light emitter and sensor are on the bottom of the mouse and you can use different types of tape to stop the light and sensor on these as well.

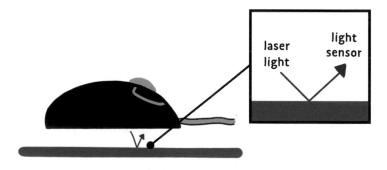

laser
light

light
sensor

PRANK REVIEW: REMOTE CONTROLLED

Be sure to document your reactions to Remote Controlled!

- What did I learn from this prank about electromagnetic light?
- What did I learn from this prank about sending signals with light?
- Why does some tape work on some light and not on others?

DR. FREEZE

 **VICTIM:** Sibling

 **FUNNINESS:** 5

 MESS: 0

 **SCIENCE:** 9

 **DANGER:** 2

SHOPPING LIST

- ☐ Super pure water (like Fiji—anything labeled "Pure" should work)
- ☐ A freezer

WARNINGS TO FUTURE ME

- Be careful when handling the cold water, any slight bump can trigger it.

- This is a difficult prank to get right; it may take a few tries.

- The less pure the water, the more likely it will freeze in the freezer. We don't want this to happen!

- The time it takes to reach the freezing point will depend on how cold and how well the freezer will cool. May have to do a few practice runs.

COLD HARD FACTS

Ice cubes from your freezer are often cloudy because they freeze quickly and contain impurities. If you freeze pure water very slowly, it is possible to make a perfectly clear ice cube!

PURE WATER

PRANK PROCEDURE:
DR. FREEZE

INSTRUCTIONS

PURE WATER

PURE WATER

1. Take your bottle of supercool water and place it **UPRIGHT** in the freezer.

2. Leave it in the freezer for ~~2 hours~~, ~~4 hours~~, ~~5 hours~~, 6 hours.

3. **You will need to continually check on it every hour.** When checking on the water, make sure it hasn't frozen but is still cold enough that it could. (Try keeping another bottle of tap water with the pure water to help you know when it is ready.)

2 HOURS LATER...

4 HOURS LATER...

6 HOURS LATER...

Works best on a really hot day.

"You want some ice-cold water?"

PURE

PURE

PURE WATER

Now they'll have to wait for their drink to thaw. HA!

INSTRUCTIONS

4. Ask your victim, "Would you like some ice-cold water?"

5. CAREFULLY take the water out of the freezer and allow your victim to see the liquid inside.

6. SLAM it down on the table in front of them.

7. The entire bottle should freeze solid before their eyes.

8. Casually walk away. (It'll look awesome.)

HOW DOES IT WORK?

Impurities in water allow ice crystals to form easily as they create *nucleation* points.

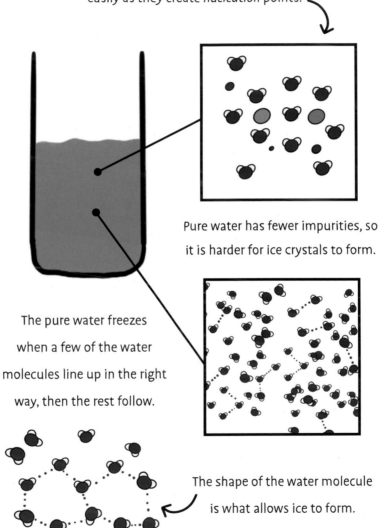

Pure water has fewer impurities, so it is harder for ice crystals to form.

The pure water freezes when a few of the water molecules line up in the right way, then the rest follow.

The shape of the water molecule is what allows ice to form.

Scientists can look at the shape of certain substances and see how the molecules and atoms are arranged by shooting a laser through a crystal containing the substance.

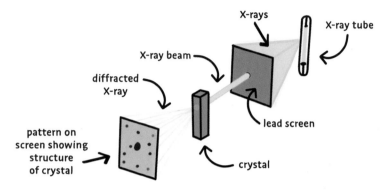

This is how we found out about the structure of DNA and the shape of some atoms!

This image can actually help scientists see the structure of a material and the way the atoms are arranged

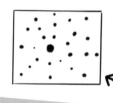

THE NEXT LEVEL: DR. FREEZE

MORE SCIENCE!

Try putting multiple bottles in the freezer in different places to see where the coldest parts are where the best place to freeze a bottle is.

You can make your own purified water by boiling tap water and collecting the steam.

PARENTAL SUPERVISION REQUIRED

BE CAREFUL as you figure out how to do this because boiling water is hot! DANGER LEVEL GOES TO 9.

FAST FACT

Water boils at much higher temperatures than other similar molecules because the oxygen and the hydrogen (H_2O) are so strongly attracted to each other.

FAST FACT

The shape of water molecules makes it a very special chemical. It is one of only a few substances where the solid form is LESS dense than the liquid form, which is why ice floats. The shape of the molecule is also the reason snowflakes always form in hexagonal (six-sided) patterns.

ASTRO FACT

The elements that make water, hydrogen and oxygen, are the first and third most abundant elements in the entire universe! We know this thanks to astrochemists. Astrochemistry is study of how much stuff is out there and what it is doing. It's chemistry...in SPACE!

PRANK REVIEW: DR. FREEZE

Be sure to document your reactions to Dr. Freeze!

- What did I learn from this prank about crystals?
- What did I learn from this prank about nucleation and how water freezes?
- How can a laser and a crystal help us figure out the structure of stuff?
- Who was Rosalind Franklin?

DIAPER GENIUS

 VICTIM: Mom

 FUNNINESS: 9

 MESS: 8

 SCIENCE: 7

 DANGER: 7

SHOPPING LIST

- ☐ Disposable diaper
- ☐ Light-colored cup
- ☐ Salt
- ☐ Scissors

WARNINGS TO FUTURE ME

- Make sure to use enough of the powder in the glass, or else it will just fall out as a soggy goop.
- DO NOT let your victim drink it.
- Dispose of it in the trash, not the sink.

PRANK PROCEDURE:
DIAPER GENIUS

INSTRUCTIONS

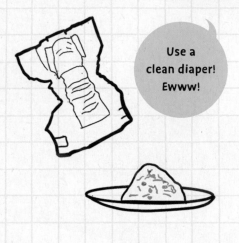

Use a clean diaper! Ewww!

1. Cut out the center part of the diaper, where the pee goes.

2. Take out the cotton-like stuff on the inside.

Shhh...

3. Hide it in the bottom of the light-colored cup. (Using a light-colored cup helps hide the white diaper stuff.)

Valeri Hunter Gordon Invented the disposable diaper out of nylon parachutes, tissue wadding, and cotton wool in 1947 after she was tired of washing her childrens' soiled reusable diapers.

4. When your victim pours liquid into the cup, the chemical will absorb it and turn into a gel.

5. The gel will stay in the cup! HA HA HA.

6. Make sure there is enough of the stuff to absorb a full cup of liquid and **DON'T LET YOUR VICTIM DRINK IT!**

WHAT'S HAPPENING?

There are two parts of the diaper that make this work:

1
THE COTTON

(makes it comfortable
on baby butts)

2
SODIUM POLYACRYLATE POWDER

(superabsorbent,
water lock)

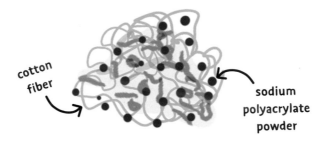

cotton
fiber

sodium
polyacrylate
powder

DID YOU KNOW?

Cotton itself is also very absorbent. Cotton is a soft fiber that grows in a little ball on cotton plants. Cotton plants were domesticated independently in both the Old World (Europe, Asia, and Africa) and the New World (North and South America) many thousands of years ago.

FAST FACT

Sodium polyacrylate can absorb about five hundred times its own weight in water.

SODIUM POLYACRYLATE

An element

A chain

Made from acrylic

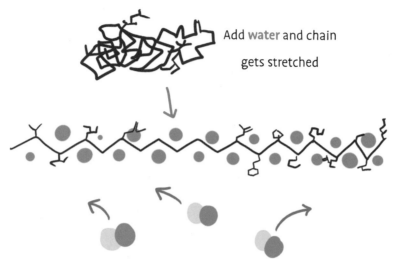

Add **water** and chain gets stretched

To turn it back into liquid, you can just add A LOT of salt! Salt will unstick the water.

FAST FACT

Disposable diapers sent to landfills can take somewhere between 250 and 550 years to decompose.

THE NEXT LEVEL: DIAPER GENIUS

MORE SCIENCE!

Industrial scientists have to test different products to see which is best. Let's do this with different brands of disposable diaper.

You will need a few different types of diapers, a measuring cup, some other cups, and some water.

1. Cut the material out of each diaper.
2. Put each diaper material into its own cup.
   * **must be the same amount of material in each cup**
3. Add some water to each cup.
   * **must be the same amount of water...say half a cup?**
4. Add small but equal amounts of water to each until one of the cups can no longer absorb water. This is the least useful diaper.
5. Keep going 'til you can rank all of the diapers from most absorbent to least absorbent.

PRO TIPS

- If it's too hard or weird to buy diapers (like if no one in the family has peed their pants in a loooong time), you can buy water storage crystals from a gardening store.

- May have to mix them into some cotton wool.

- Will have to find the right balance of cotton wool and crystals.

- Picking the perfect diaper for this prank may take some experimenting.

SCIENCE FACT

Scientists strive for repeatability in their experiments. For an experiment to be more repeatable, you need:

- the same procedure
- the same conditions
- the same measuring equipment
- the same location

The same sort of chemicals can be used to make fake snow!

PRANK REVIEW: DIAPER GENIUS

Be sure to document your reactions to Diaper Genius!

- What did I learn from this prank about absorbency?

- What did I learn from this prank about using chemicals to help hold on to water?

- Why do diapers, gardening crystals, and fake snow all need to hang on to water molecules?

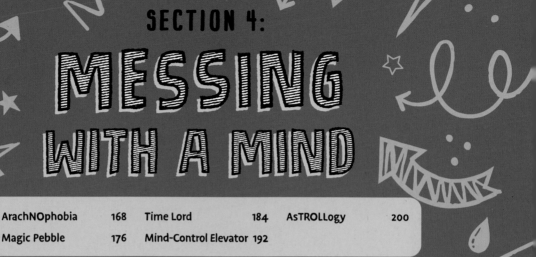

SECTION 4:
MESSING
WITH A MIND

While all of the pranks in this book are designed to make the victim question what happened to them, this section has pranks specially designed to mystify and confuse the victim into believing in nonsense or acting out of character. Be warned, sometimes making people feel foolish can bring you into the firing line for the next prank, so keep it in good spirits.

To make these pranks work, you need to not only say the right things, but also seem like you are thinking the same thing they are. It helps if you understand a bit about psychology, but we've put some clues in the pranks to help you along the way. Remember, you can't trick someone if you believe it's not a trick. As soon as you smirk, smile, or show that you know what is wrong, they will figure it out and the jig is up. Use the "Prank Review" to nail down your act and become a master mentalist.

EXPERIMENT #21

ARACHNOPHOBIA

VICTIM: Anyone who hates spiders

FUNNINESS: 8

SCIENCE: 6

MESS: 2

DANGER: 6
Ask an adult first

SHOPPING LIST

- ☐ Scissors
- ☐ Thick paper (or cardboard)—Dark colors work best
- ☐ Double-sided tape
- ☐ Lamp with shade

WARNINGS TO FUTURE ME

- DO NOT put the cutout on the light—it can heat up and become a fire hazard.
- Remove the cutout from the lamp once the prank is completed.
- DON'T do it to Grandma because she may have a heart attack!
- Be sure to film your victim's reaction.

DID YOU KNOW?

The word "arachnophobia" comes from two words: Arachnid—animal group including spiders and scorpions named after Arachne, a weaver from Greek mythology; and phobia—meaning "fear of." "Arachnophobia" literally means a fear of spiders!

PRANK PROCEDURE:
ARACH**NO**PHOBIA

INSTRUCTIONS

Trace over this one if you're stuck.

1. Draw a spider on a piece of paper. Make sure it is thick enough to make a good silhouette.

2. Carefully cut out the shape of the spider. The better the cutout, the more realistic the shadow on the lamp will appear.

3. Pick which lamp to stick the spider to and unplug the lamp from the power socket.

4. Use CLEAR sticky tape to secure the spider to the INSIDE of the lampshade.

INSTRUCTIONS

5. Rotate the lamp so the spider is right in front of your victim.

6. Plug the lamp back in and switch it on to test if your silhouette works and looks scary.

7. Turn off the lamp and wait for your victim to sit down and turn on the light themselves. You may want to darken the room by closing blinds or curtains to encourage them to turn on the lamp.

→ SOME NOTES ←

☑ Arachnophobia is not just a fear of spiders but of scorpions as well.

☑ Arachnids are not insects. They have eight legs and two main body segments, while insects have six legs and three main body segments.

☑ Other arachnids include mites and ticks. They all use venom and can make people very sick, which is why people are naturally afraid of spiders!

☑ The majority of spiders are harmless. In fact, spiders are extremely important in maintaining insect populations that help protect farmers' crops.

SPIDER SCIENCE

The golden orb spider's silk that is used to make its web is actually stronger than steel. We don't notice this because the silk in the web is so thin.

DID YOU KNOW?

It is widely thought that the movie *Jaws* led to widespread fear of sharks (or selachophobia) in the 1970s.

FAST FACT

The black widow spider gets its name from the fact that the female spiders engage in "sexual cannibalism." This is when, after mating, the female will eat the male. Hence the term "widow" in the name.

There are many different types of phobias.

☑ Many people are scared of animals other than spiders, like snakes (ophidiophobia) or even dogs (cynophobia).

☑ People can also have phobias for other things like crowded spaces (agoraphobia) or heights (acrophobia).

☑ Phobias can develop from a person's experiences. They may have had a bad time once that left them permanently scared of something. Or they can develop phobias from seeing something scary!

☑ Some people may have phobias from birth, suggesting it is in their DNA!

THE NEXT LEVEL: ARACH**NO**PHOBIA

MORE SCIENCE!

People can have all kinds of reactions to different animals depending on their attitudes to them.

1

You could experiment with this idea by tracing and cutting out silhouettes of some other animals like beetles, lizards, or even spiders of different sizes and shapes.

2

You can examine the different reactions to build a rundown of the scariest animals for your friends and family.

3

FAST FACT

You can tell the difference between arachnids and insects by the numbers. Insects have six legs, two body segments, and at least two wings; arachnids have eight legs, two body segments, and no wings.

⋛ FAST FACT ⋛

One of the world's largest spiders is the giant huntsman. It can have a leg span of up to twelve inches—about the size of a dinner plate!

BRAIN SCIENCE

There is no universal cure for a phobia. One type of treatment is cognitive behavioral therapy, where patients are slowly exposed to their phobia and taught ways to cope. Sometimes this can even start with virtual reality simulations.

CAN YOU SPOT IT?

The world's smallest spider has never been seen. The female Patu digua spider has a body length of less than 0.015 inches. Since males are almost always smaller than females, the male Patu digua spider should be very small indeed—so small that no one has ever seen one!

PRANK REVIEW: ARACH**NO**PHOBIA

Be sure to document your reactions to ArachNOphobia!

- What did I learn from this prank about phobias?
- What did I learn from this prank about how people react to spiders?
- Is there an advantage to having a phobia?
- How could a person overcome a phobia?

MAGIC PEBBLE

VICTIM: Gullible friend

FUNNINESS: 8

SCIENCE: 7

MESS: 0

DANGER: 5 might throw pebble at you

SHOPPING LIST

☐ Pebble or small object (A crystal or something shiny will be more convincing)

WARNINGS TO FUTURE ME

- Do not push down too hard on the victim—slowly increase the force of your push until the desired effect is achieved.

- Get the victim to do each of the experiments only once. The more times you do it, the more likely they are to catch on to what you're doing.

- You will need to practice this prank and exactly what to say.

- The "pebble" is only for effect, you could really use anything—a special card, bracelet, stickers, whatever! But the item should look special or unique for best effect.

- Try to use geek-speak to make it sound more technical and believable; words like "frequency," "resonance," "quantum," and "alignment" are awesome!

Maybe say something like, "Oh wow! Do you know what kind of rock this is? The molecules in this pebble resonate at a frequency similar to the molecules in your body. It can help with the alignment of your muscles and your nerves at a quantum level."

PRANK PROCEDURE:
MAGIC PEBBLE

INSTRUCTIONS

AD FACTS

Buzzwords are used in advertising and marketing to confuse and trick people into buying certain products.

1. Come up with the perfect description of the magical and scientific properties of your pebble, and run through it with your victim. If they don't believe you, that's fine! You now must say "let me prove it to you."

2. Have the victim stand on one leg with their hands outstretched.

3. Place both hands on their shoulders and push down gently on an angle slightly TOWARD the side with the raised leg—it should put them off balance.

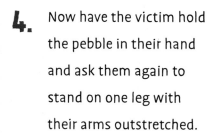

4. Now have the victim hold the pebble in their hand and ask them again to stand on one leg with their arms outstretched.

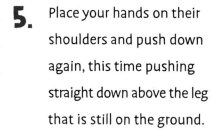

5. Place your hands on their shoulders and push down again, this time pushing straight down above the leg that is still on the ground.

6. They should find their balance has improved and it is harder to make them fall down with the pebble! WOW!

CRYSTAL CLEAR

Worldwide sales of crystals amounts to more than $1 billion per year! Unfortunately, if you buy them hoping for a magical effect, you are going to be disappointed! Magic is great fun to read about and see on screen, but it is not reality.

THE NEXT LEVEL: MAGIC PEBBLE

VICTIM STILL NOT CONVINCED?

1. You can tell them you know another test that will again prove the magic of the pebble!

2. Without the pebble, have your victim stand with their feet shoulder-width apart and get them to stick their thumb straight up as if they were giving you a thumbs-up.

NOTE

You could do something to it, like hold it up to the Sun to "energize" it and really sell the prank.

⸪ FAST FACT ⸪

The Sun is said to be the source of all energy on Earth. Indeed, holding your hand and pebble out in the sunshine will transfer energy between the sun and the pebble. But, that energy will be thermal energy that works only to heat up the pebble, and nothing more.

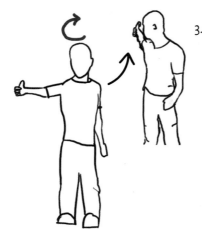

3. Without moving their feet, have your victim rotate the top part of their body as far around as they comfortably can with their thumb still outstretched. Ask them to be sure to notice how far around they went by looking at what their thumb was covering in the background.

4. Now give your friend the pebble.

5. Ask them to repeat the test, keeping their feet still and rotating as far as they can comfortably go.

6. Encourage them to go as far as they can and on the second attempt they should have gone farther! Magic!

⊰ DID YOU KNOW? ⊱

Scientists regularly observe the placebo effect, which is a strange happening where people who believe something will help them actually perform better than people who don't, even though there is no real difference between them. Another effect called the nocebo effect works the other way where people who believe something will hinder them perform worse than people who don't.

✏️ ➡ WHAT'S HAPPENING? ⬅ ✏️

This prank works because of a combination of psychology and physics. The direction you push down on your victim's shoulders helps convince them the pebble has improved their balance.

We often use arrows to indicate the direction of force on an object and a force applied in a slightly different direction can have very different results on the object.

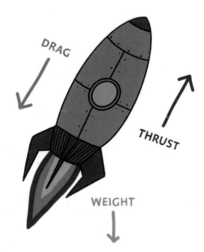

DRAG

THRUST

WEIGHT

Muscle memory also helps convince the victim that the pebble is helping them reach farther. Once your body has done something once, it finds it easier to do it again. The person would reach farther whether they had the pebble or not!

DID YOU KNOW?

The pranks here have actually been used to trick people into buying certain products. A business was claiming that their bracelets had holographic technology that would help improve things like balance, strength, and flexibility. The demonstrations that were conducted to convince people were very similar to the pranks you're playing on your victim.

DID YOU KNOW?

Psychology is the scientific study of human behavior and mental processes. It began as an experimental science in 1870s Germany.

PSYCH OUT

Many of the world's top athletes wear the same clothing or jewelry, or have rituals before they compete because they believe it will improve their performance. While it's unlikely any of these rituals will give any physical advantage, sometimes performance is a mental thing. This is called the "placebo effect."

PRANK REVIEW: MAGIC PEBBLE

Be sure to document your reactions to Magic Pebble!

- What did I learn from this prank about applied forces and direction?
- What did I learn from this prank about people's psychology and the placebo effect?
- Why does our body move so much more freely after just one attempt?

EXPERIMENT #23

TIME LORD

VICTIM: Dad

FUNNINESS: 7

MESS: 0

SCIENCE: 8

DANGER: 2

SHOPPING LIST

- ☐ Strong, thin magnet (you can use the same rare earth or neodymium magnet from the Anti-Coaster prank!)
- ☐ Quartz analog watch

WARNINGS TO FUTURE ME

- Strong magnets are...well, STRONG!
  - → can pinch you
  - → can damage things
  - → can ruin electronics
- Make sure to use a quartz analog watch, as neodymium magnets will break a mechanical watch.
- Use your own watch. Don't use Dad's expensive watches!

neodymium
magnet

INSTRUCTIONS

The watch will not work!

1. Attach the magnet to the back of the watch.

"Oh no! My watch isn't working! Please help me!"

2. Tell your victim "Oh no! My watch isn't working! Please help me!" Be sure to show them time is frozen before you hand over the watch.

3. Sneakily remove the magnet as you take off your watch for your victim to take a closer look.

DID YOU KNOW?

Before clocks and watches, people would get the time from the local church that would chime its bells during the day. This was notoriously inaccurate and even villages close to each other would often work under different times.

DID YOU KNOW?

The first quartz clock was made in 1927 at Bell Telephone Laboratories, however, it wasn't until 1969 in Japan that the first quartz wristwatch was made.

INSTRUCTIONS

"Now it's working fine!"

4. The victim should be confused, and say something like, "Now it's working fine!"

5. Put it back on your wrist, and, after sneakily attaching the magnet again, say, "Hey! It's stopped again!"

DID YOU KNOW?

A day is made up of twenty-four hours because of ancient Egyptians. They divided the night into twelve parts and the day into another twelve parts based on the position of certain stars in the sky.

6. And the cycle continues...

→ WHAT'S HAPPENING? ←

This prank works because of the parts inside a watch.

REALLY FAST FACT

Quartz vibrates depending on how it is cut. In most watches, the vibrations happen 32,768 times per second.

The battery powers the quartz, the quartz vibrates consistently, the computer chip counts the vibrations and tells the motor to move, the electromagnetic motor moves the gears, and the gears move the watch hands.

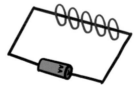

The electricity made by the battery moves in a loop creating a magnet. Putting another magnet inside the watch and then attaching gears to it makes a motor!

The motor and the gears drive the hands on the watch face as long as the electricity from the battery is flowing.

Sticking a magnet on the outside of the watch disrupts the motor and stops the hands from moving.

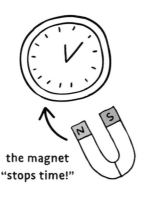

the magnet "stops time!"

FAST FACT

Electricity and magnetism were once thought to be different but are now considered the same force and together are called "electromagnetism."

DID YOU KNOW?

The ancient Babylonians are responsible for us having sixty-minute hours and sixty-second minutes, though no one knows for sure why sixty was used. Maybe because sixty is a multiple of so many numbers (one, two, three, four, five, six, ten, twelve, fifteen, thirty, sixty)?

MAKING YOUR OWN MOTOR

You will need to get a AA battery, your magnet, and some copper wire.

Attach the magnet to the "–" end of the battery

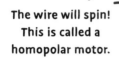

The wire will spin! This is called a homopolar motor.

Bend the wire so the middle of it can touch the "+" end of the battery, before folding it up and then back down so each end of the wire touches the battery at the "–" end of the battery

DID YOU KNOW?

In 2018, scientists at Purdue University made nanoparticles of silica spin 10 billion times per second using lasers! This is fastest spinning object made by humans.

PRO TIP

Sometimes the copper wire is covered in a special material (called an insulator) that prevents the wire from spinning. Using some ordinary sandpaper to scrape off the layer will allow it to work more efficiently.

DID YOU KNOW?

Quartz is a crystal made up of oxygen and silicon atoms. It's found in many countries around the world, as it is one of the most abundant minerals in Earth's crust.

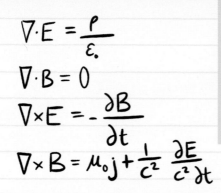

$$\nabla \cdot E = \frac{\rho}{\varepsilon_0}$$

$$\nabla \cdot B = 0$$

$$\nabla \times E = -\frac{\partial B}{\partial t}$$

$$\nabla \times B = \mu_0 j + \frac{1}{c^2} \frac{\partial E}{c^2 \partial t}$$

James Clerk Maxwell's calculations found that light was an electromagnetic wave. Like other electromagnetic waves, it could travel at 300,000,000 meters in a single second!

PRANK REVIEW: TIME LORD

Be sure to document your reactions to Time Lord!

- What did I learn from this prank about how a quartz analog watch works?
- What did I learn from this prank about how electromagnetism can be used to make a motor?
- What would happen to the hands of a watch (or the copper wire) if the magnet in the motor is flipped over?

EXPERIMENT #24

MIND-CONTROL ELEVATOR

VICTIM: Strangers

MESS: 0

DANGER: 3

FUNNINESS: 6

SCIENCE: 6

SHOPPING LIST

- ☐ At least 4 friends (works better with one or more adults)
- ☐ A mall elevator
- ☐ A straight face

WARNINGS TO FUTURE ME

- DO NOT do this prank alone. You will need at least one friend with you at all times to watch your back and to keep you from looking like a complete weirdo.
- The more people in this prank, the higher the chance of success.
- Be careful with this prank, because it involves people you do not know. Be respectful. (For example, if you decide to film this prank, do not post or show the video to anyone without the permission of everyone in the video.)

PRANK PROCEDURE:
MIND-CONTROL ELEVATOR

INSTRUCTIONS

1. Get in the elevator with your co-pranksters, making sure everyone knows how this should work and that there are no other people in there when you set it up.

2. Select the floor that will take you on the longest journey in the elevator. (Don't press all the buttons—no one likes that person.)

3. Spread out all of your accomplices so it looks like you don't know each other. It looks better if two people are together but everyone else is separate.

DID YOU KNOW?

Elevators in some form or another have been around for many hundreds of years. However, it was only in 1852 that Elisha Graves Otis invented the safety break and made them reliable and safe. The Otis Elevator Company is still one of the largest manufacturers of elevators today.

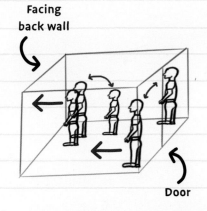

Facing
back wall

Door

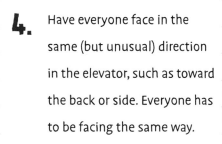
4. Have everyone face in the same (but unusual) direction in the elevator, such as toward the back or side. Everyone has to be facing the same way.

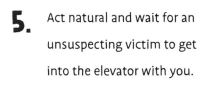

5. Act natural and wait for an unsuspecting victim to get into the elevator with you.

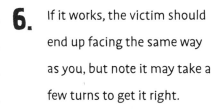

6. If it works, the victim should end up facing the same way as you, but note it may take a few turns to get it right.

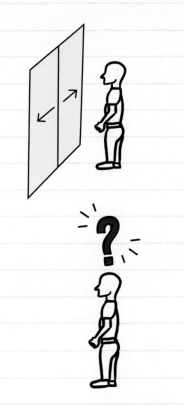

7. Try facing in different directions and record how many minds you end up controlling.

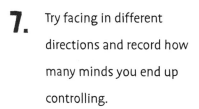

WHAT'S HAPPENING?

☑ This prank is based on a series of psychology studies from the 1960s that found that humans are susceptible to something known as groupthink.

☑ Humans desire to be accepted by other people, and our subconscious mind influences our behavior to ensure we are conforming with others, whether we know them or not!

FAST FACT

Scientists have found people will regularly give up their own beliefs and even what they think is right or wrong if it allows them to fit in with a bigger group!

HOT TIP

Too much cooperation can be a bad thing. The science of successful teams shows that a balance must exist between cooperation and conflict. The next time you build a team, make sure there is room for a wide range of ideas.

Humans are known for making poor decisions because of groupthink.

Watch this!

People often drive too fast in order to look cool in front of their friends. We see an increase in the number of fatalities and risk-taking behavior when young drivers have passengers.

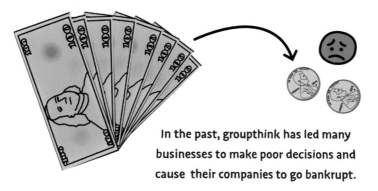

In the past, groupthink has led many businesses to make poor decisions and cause their companies to go bankrupt.

Kids often talk or act a certain way when they're around "cool kids" they want to impress, even if they wouldn't talk or act that way normally. If you have to do certain things you don't like in order to fit in, you don't want those kids as your friends in the long run.

THE NEXT LEVEL: MIND-CONTROL ELEVATOR

MORE SCIENCE!

☑ Try to change your positions in the elevator so the victim is in the middle, in front of, or behind you to figure out which works best.

☑ You could test similar ideas about groupthink in waiting rooms or in crowds. If you and your friends feel brave, try singing a well-known song (like "Happy Birthday") in a crowd and see how many people join in.

☑ The prank only works if the victim feels like they are part of the group. If it is obvious to the victim that everyone in the prank knows each other, it will not work.

QUICK QUIZ

Is science the universal cure? Science is practiced by scientists, who are also humans, of course. Could science itself be susceptible to groupthink?

DID YOU KNOW?

In 1999, an amazing display of groupthink occurred when all the Major League Baseball umpires resigned together. They believed the group decision would allow them more negotiating power with the sports organization. It did not.

DID YOU KNOW?

Psychologists believe that the best way to avoid falling victim to groupthink is to do the following:

→ Allow a full discussion to take place so that everyone is aware of the issue.

→ Speak up when you think something is incorrect or if you have a differing view.

PRANK REVIEW: MIND-CONTROL ELEVATOR

Be sure to document your reactions to Mind-Control Elevator!

● What did I learn from this prank about influencing a stranger?

● Why are humans so susceptible to groupthink?

● Why does it only work some of the time?

● How can groupthink be used for good and evil?

EXPERIMENT #25

AsTROLLogy

VICTIM: Anyone

FUNNINESS: 9

MESS: 1

SCIENCE: 8

DANGER: 2

SHOPPING LIST

☐ 6 envelopes

☐ 6 identical horoscope readings

☐ 6 family members and/or friends

☐ printer

WARNINGS TO FUTURE ME

- Some people really do believe in this stuff. Be nice to them and make sure they know you're just having fun!

- Astrology is the study of how distant celestial objects such as planets and stars affect our personal lives... Spoiler alert...they don't.

- Astrologers believe that the arrangement of the Sun, Moon, stars, and planets at the time of our birth influences who we are as people. There is very little scientific evidence to back this up (or none).

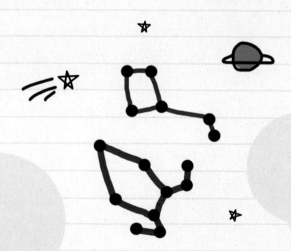

PRANK PROCEDURE:
AsTROLLogy

INSTRUCTIONS

1. There is some planning involved in this prank and it works best with a group of people, so you can prank more people at once!

2. Ask your victims to write down their star sign (or zodiac sign) on a piece of paper and give it to you.

3. Go online and choose one horoscope from any astrology website.

It doesn't matter which horoscope or website.

Don't let anyone see anyone else's "horoscope."

← REMEMBER, you should have the same reading for each victim, just a different title at the top.

INSTRUCTIONS

4. Copy the text and print out the same horoscope for each of your victims, changing the star sign at the top of the horoscope to their specific star sign.

PRO TIP: Make sure to read through the horoscope to change any mention of star signs in the text to the victim's specific sign.

Not even close. Kind of... I guess. Wow!

0 5 10

5. Return the horoscope to each person and ask them to rate how accurate their horoscope is.

The purpose of the readings is to make them relevant for a large group of people. The problem is anyone can connect to vague statements. "You will experience periods of sadness" relates to everyone!

6. Compare the results with each other before revealing they were all the same horoscope.

→ WHAT'S HAPPENING? ←

This prank works because many people have shared experiences, like how they feel and what they do on a daily basis. The zodiac sign just makes it more believable because it is a gimmick.

If you wake up super early on your birthday and watch the sunrise, you will be able to see which of the twelve zodiac signs is in the middle of the sky

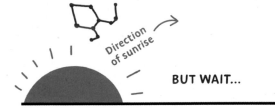

BUT WAIT...

The newspaper will say your star sign is something else. This is because over two thousand years ago when star signs were created, the Earth was leaning on a different angle and has since wobbled, so everyone is actually off by a whole month!

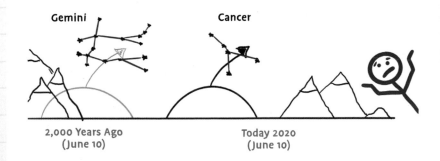

THE NEXT LEVEL: AsTROLLOGY

MORE SCIENCE!

To really take your mystical knowledge up a notch,
you could try out some "cold reading."

Psychics, fortune-tellers, and mediums use
cold reading to get a lot of information about
someone with minimal effort. Some even
claim they can communicate with the dead!

DID YOU KNOW?

Astrology originated over 3,000 years
ago in Babylon and spread to the
Mediterranean about 2,400 years ago.

FAST FACT

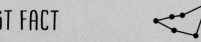

There are different types of astrology. "Sun sign
astrology" is based on the position of the sun at the
time of your birth within one of the twelve zodiac signs,
which is then called your "sun sign" or "star sign."

SCIENCE FACT

People believe in astrology and mind readers because of the Barnum effect. This occurs when broad general descriptions are mistakenly interpreted as highly personal. e.g. "I see someone close to you who likes wearing a black coat."

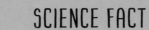

Here are some tips about how to cold read, but to do it well it takes a ton of practice.

→ Speak with confidence because it makes you more believable.

→ Use gimmicks, like crystal balls, tarot cards, or gemstones—anything that seems supernatural.

→ Know your audience. Research and remember small details about your victim from what they or their family and friends have said.

→ Be modest, and remember you are just there to "help" them.

→ Always be vague and try to give the impression that you know more than you are letting on. Say things like, "I get the sense that there is something important about you." "I feel like there is more to your story than meets the eye."

→ Get information from the victim by rephrasing statements as questions and retelling what they say back to them another way. "Was your father a good cook?" "Yeah, he made dinner for the family every night." "Yes, I'm seeing an image of him working in a kitchen..."

→ Always flatter the subject.

→ If things don't go as planned, have an escape route. ⟶

"Were your parents teachers?"

A vague statement that could be considered a very specific question.

"No. My mother was a mechanic."

"But she would always be teaching you about cars."

Parents always teach their children how to do things.

"OMG! She would spend hours showing me how to change tires."

"She was a very hard worker, wasn't she?"

Flattering statement

PRANK REVIEW: AsTROLLogy

Be sure to document your reactions to AsTROLLogy:

- What did I learn from this prank about astronomy vs. astrology?
- What did I learn from this prank about useful statements for pranking people?
- Why is vagueness so important?
- How should a "psychic" act to convince people they are telling the truth?

GLOSSARY

ACID: Any substance that has a pH value lower than 7. Acids are corrosive and have a sour taste. Something is considered an acid if it has free-moving protons. The more free-moving protons, the stronger the acid and the lower the pH value.

AIR PRESSURE: The force exerted on an object by the weight of the air around that object. The closer you are to sea level, the greater the air pressure. Air pressure is measured in *atmospheres* (atm) or kilopascals (kPa). 1 atm is equal to the average air pressure at sea level, which is about 100 kPa.

ARACHNIDS: Hard-bodied animals with eight legs and two main body parts (the head and abdomen). This group of animals includes spiders and scorpions.

ATMOSPHERE: The mixture of gases covering the Earth. It is made up of 78 percent nitrogen, 21 percent oxygen, a bit of argon, and other very small amounts of a large number of other gases like carbon dioxide and ozone. Water vapor also makes up a part of the atmosphere but varies widely at different elevations and is sensitive to variations in temperature and air pressure.

ATOMS: Tiny particles that combine to make up everything that has mass, including you and me. Atoms are made up of even smaller parts called "subatomic particles." Subatomic particles include protons and neutrons, which can be found stuck together at the very center of atoms in a place called the "nucleus." Then there are electrons, which move around the outside of atoms in specific places called shells. There are many different types of atoms called elements.

BACTERIA: Super small living things that are a relatively basic cells made up of a cell wall and membrane, some important molecules, and a jellylike fluid.

BASE: Any substance with a pH value greater than 7. Bases can be used in cleaning products and taste bitter. Something is considered a base if it is able to remove protons from a substance, even taking them away from other molecules. The better the ability of the base to remove protons, the stronger the base and the higher the pH value.

CELL: The smallest part of any living thing. There are many different kinds of cells, ranging from the relatively small and simple to the larger and more complex. Cells do specific jobs that help keep things alive. Sometimes an individual living thing might be made of only one cell, like a bacteria, or a living thing could be made of billions and billions of them, like you and me or plants and other animals.

CONDUCTOR: Any material that allows heat or electricity to easily move through it. Good heat conductors heat up and cool down quickly. Good electrical conductors have free-moving particles (called electrons) that can carry an electric charge.

DENSITY: The number of particles or amount of mass in a given space. Substances with greater density have more mass in a smaller area. If something is less dense than water, it will float; if it is more dense, it will sink.

Heavy ⟶   ⟵ Light

DIGESTION: The body's way of getting access to nutrients by breaking down food into smaller parts. This can happen through mechanical means, like chewing or churning in the stomach, or chemical means, like stomach acid dissolving food or enzymes breaking down food molecules.

ELEMENT: When atoms have the same number of protons, they are said to be the same element. For example, every single oxygen atom has 8 protons and each nitrogen only has 7 protons. If you could add another proton to a nitrogen atom, it would no longer be nitrogen, it would be oxygen. Even though oxygen and nitrogen are both atoms, they are different elements. All of the elements we know about are arranged in a table called the periodic table of elements.

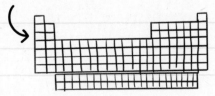

ENERGY: There is no time involved in measurement of force. Energy exists in many forms and can be transformed between them, for example chemical, kinetic, nuclear, elastic etc... You can never "lose" energy, it can only be changed from one kind into another. For example, when you switch on a light in your house, the electrical energy flowing through the wires in your house changes into light energy and heat energy in the light bulb.

ENZYME: Small molecular machines that work inside cells and throughout your body. They have very specific jobs to do and they make biological processes more efficient by reducing the energy required to complete the process.

EUSTACHIAN TUBE: A small tube connecting the ear with the sinus, used to control the pressure in the head and help with hearing more clearly in different environments.

EVOLUTION: The gradual changing of one species to another to better survive in certain environments. As an environment changes, certain living things may struggle to survive, while others may thrive. The ones that are better suited to the environment make more babies, making their population and their particular characteristics flourish. On the other hand, the ones that struggle have fewer babies and slowly decline, maybe even go extinct.

EXTINCT: When a type of life-form no longer exists in an area, it is said to be extinct.

FORCE: A push, pull, or twist experienced by an object that may change its speed, shape, or direction of travel. It is measured in Newtons. One Newton is equal to about the force required to lift four AA batteries about three feet off the ground in one second.

FREQUENCY: The number of waves passing a specific point every second or how often a certain event happens in a specified time. It is measured in hertz (Hz). An object with a frequency of 50 Hz repeats fifty times in one second.

GASES: Free-moving particles with no fixed shape or volume.

HEAT: Thermal energy transferred between things to change their temperature. Heat transfers from hot things to cold things without effort. Making things colder (like a freezer) requires extra energy.

HYPOTHESIS: An idea or prediction to explain an observation that can be tested in an experiment. A hypothesis can be either accepted or rejected. For example, "I think the sky is blue because of light reflecting off the blue ocean." (This is NOT the reason the sky is blue. This means your hypothesis should be rejected.)

INSULATOR: Any material that does not allow heat or electricity to freely move through it. It is the opposite of a conductor.

LIQUID: Free-moving particles with no fixed shape, but with a specific volume.

MEMBRANE: A thin, flexible covering that allows some substances to pass across it. All cells have a membrane separating the jelly stuff inside from the other stuff on the outside.

MICROBE: An extremely small life-form that can only be seen through a microscope.

MIXTURE: A bunch of different stuff combined into a single substance, easily separated.

 **MOLECULE:** A number of atoms joined together by chemical bonds, difficult to separate.

NATURAL SELECTION: The process by which evolution occurs. It helps ensure the survival and reproduction of life-forms that are best suited to the environment in which they live.

NOCEBO EFFECT: When people who believe something will hinder them perform worse than people who don't, such as if a fake treatment worsens a person's condition simply because they expect it will.

NUCLEATION: A process by which a small number of particles creates a pattern that more and more particles are able to attach themselves.

PARADOX: An idea or statement that is true, but that seems to contradict itself. For example, the grandfather paradox says that if you could travel back in time to kill your grandfather, then you would never be born, so then how could you kill your grandfather? Confused?

PH: The measure of how acidic or basic a substance is. It is a measure of the number of free protons in a substance. A pH below 7 indicates an acid. The lower the value, the stronger the acid. A pH above 7 indicates a base. The higher the value the stronger the base. A pH of 7 exactly is called "neutral" and indicates a perfect balance between the number of free protons and the number of molecules that can attach to a proton. Pure water has a pH of 7.

PLACEBO EFFECT: When people who believe something will make them perform or feel better, even though the medicine or treatment does not actually do anything to treat their condition.

PRESSURE: The amount of force experienced within a certain area. A lot of force experienced in a smaller area means there is high pressure, and little force, spread out over a greater area, means there is low pressure.

Pumped-up basketball has more air pressure

Flat basketball has very little air pressure

PROBABILITY: A measure of the likelihood of an event occurring, where "0" equals no chance of it occurring, "1" equals a certainty that an event will occur, and "0.5" is an equal chance of two things occurring, such as heads or tails in a coin toss.

SOLIDS: Particles that are fixed in a certain shape with a specific volume.

THERMODYNAMICS: The science that studies the relationship between temperature, movement, and energy.

VIRUS: Viruses are really small germs that can infect plants and animals (including humans) and make them very sick and behave differently from how they otherwise would.

VOLUME: The amount of space occupied by a certain substance. It is measured in gallons, liters, or pints (or a lot of ways). One liter is equal to 1000 cubic centimeters or about a quart of milk. Depending on where you live, volume is measured in many different ways, such as gallons, liters, or pints.

WAVELENGTH: The wavelength is a measurement of distance between the same point on two different successive waves.

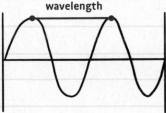

wavelength